Supply Chain Integration Challenges
in Commercial Aerospace

Klaus Richter • Johannes Walther
Editors

Supply Chain Integration Challenges in Commercial Aerospace

A Comprehensive Perspective on the Aviation Value Chain

Springer

Editors
Klaus Richter
Airbus Group
Blagnac Cedex, France

Johannes Walther
IPM AG
Hannover, Germany

ISBN 978-3-319-46154-0   ISBN 978-3-319-46155-7 (eBook)
DOI 10.1007/978-3-319-46155-7

Library of Congress Control Number: 2016961240

© Springer International Publishing AG 2017

This work is subject to copyright. All rights are reserved by the Publisher, whether the whole or part of the material is concerned, specifically the rights of translation, reprinting, reuse of illustrations, recitation, broadcasting, reproduction on microfilms or in any other physical way, and transmission or information storage and retrieval, electronic adaptation, computer software, or by similar or dissimilar methodology now known or hereafter developed.

The use of general descriptive names, registered names, trademarks, service marks, etc. in this publication does not imply, even in the absence of a specific statement, that such names are exempt from the relevant protective laws and regulations and therefore free for general use.

The publisher, the authors and the editors are safe to assume that the advice and information in this book are believed to be true and accurate at the date of publication. Neither the publisher nor the authors or the editors give a warranty, express or implied, with respect to the material contained herein or for any errors or omissions that may have been made.

Printed on acid-free paper

This Springer imprint is published by Springer Nature
The registered company is Springer International Publishing AG
The registered company address is: Gewerbestrasse 11, 6330 Cham, Switzerland

Contents

Introduction: Supply Chain Integration Challenges in the Commercial
Aviation Industry.. 1
Klaus Richter and Nils Witt

Part I Product Development

Aviation's Future Is as Bright as Its Past........................ 19
Andrew Gordon

Cabin Densification: SpaceFlex2 and Beyond........................ 31
Scott Savian

Innovation Challenges in the High-Tech, Long-Cycle Jet Engine
Business.. 43
Alan H. Epstein

Open Innovation in the Aviation Sector............................ 57
Johannes Walther and Daniel Wäldchen

Disruptive Innovation Through 3D Printing......................... 73
Reinhart Poprawe, Christian Hinke, Wilhelm Meiners, Johannes Schrage,
Sebastian Bremen, Jeroen Risse, and Simon Merkt

Part II Configuration and Demand

Fulfil Customer Order Process: Customization of Commercial
Aircraft.. 91
Gabriel Oehme

End-to-End Demand Management for the Aerospace Industry.......... 105
Avinash Goré and Alexander Nathaus

Main Differences and Commonalities Between the Aircraft and the
Automotive Industry.. 119
Horst Wildemann and Florian Hojak

Part III Component Manufacturing

Trends in the Commercial Aerospace Industry 141
Gernot Strube, Karel Eloot, Nadine Griessmann, Rajat Dhawan,
and Sree Ramaswamy

**Success Through Customer Co-Development, Global Footprint and the
Processes In-Line with the Customer** 161
Mark C. Hiller and Joachim Ley

Vertical Integration: Titanium Products for the Aircraft Industry 177
Oleg Leder

Part IV Assembly and Integration

Quality Gates 193
Isabelle Sciannamea

Lean Complexity Through Tailored Business Streams 209
Richard Hauser, Hans-Jörg Kutschera, and Benoit Romac

Driving the Digital Enterprise in the Aerospace Industry 221
Helmuth Ludwig and Alastair Orchard

Part V Life Cycle Business Models and Aftermarket

The Aero-Engine Business Model: Rolls-Royce's Perspective 237
Peter Johnston

The Material Value Chain Services in Commercial Aviation 249
Jörg Rissiek and Mikkel Bardram

**Predictive Maintenance: How Big Data Analysis Can Improve
Maintenance** 267
Jim Daily and Jeff Peterson

Outlook 279
Stefan Berndes

Introduction: Supply Chain Integration Challenges in the Commercial Aviation Industry

Klaus Richter and Nils Witt

Abstract Airlines, aircraft OEMs and their suppliers were affected by numerous changes in the commercial aviation sector over the last decades.

Since the 1970s, airlines have seen a significant increase in air traffic and seat mile cost pressure due to regulatory changes, shift of traffic towards developing countries, especially Asia, and emergence of new business models such as Low Cost Carriers.

On the OEM and supplier side, the business focus changed from defence to civil and the manufacturing base went through a phase of consolidation. Furthermore, increased demand for new aircraft and cost pressure from the airlines triggered a transition from job shop production to small series production.

This development fostered associated changes in aircraft technology such as enhanced cabin space efficiency and new engine technology. It requires a supply chain that is able to cope with the challenges of airline specific customization, while delivering into a serial production process. Cost pressure and competition from emerging countries have led to a shift in the manufacturing base, while the assembly processes have seen the adaptation of automotive concepts such as Lean Production. Additionally, the aftermarket today offers new business opportunities and will potentially benefit from revolutionary concepts in the near future due to increased use of digitalization technology.

1 Development of the Aviation Industry Over the Last Decades

The entire commercial aviation industry has gone through a series of significant changes in the last half century affecting airlines, aircraft OEMs and suppliers.

K. Richter • N. Witt (✉)
Airbus, Blagnac Cedex, France
e-mail: nils.witt@airbus.com

© Springer International Publishing AG 2017
K. Richter, J. Walther (eds.), *Supply Chain Integration Challenges in Commercial Aerospace*, DOI 10.1007/978-3-319-46155-7_1

1.1 Airlines

On the airline side, the business model evolved from being driven by tight authority regulation and state-owned flag carriers into a model based on fierce international competition.

In the 1970s, the U.S. and Europe were by far the most important markets in terms of revenue passenger kilometers and accounted together for 76 % of world airline traffic (including routes to Canada and Japan). At the same time, these two markets were tightly regulated (Airbus GMF 2011).

In the U.S., for instance, prior to the Airline Deregulation Act of 1978, fares on domestic routes were set by the government agency Civil Aeronautics Board (CAB) and airlines were only allowed to operate routes after prior approval. The only U.S. airline, for example, that was granted permission to operate transatlantic flights to Europe was PanAM. This only changed in 1979 when the International Air Transport Competition Act also allowed American competitors to offer international routes (Pompl 2007).

At the same time, in European countries many state-owned so called "flag-carriers" existed. Major players such as Air France, Lufthansa and British Airways had controlling public ownership until 1984, when British Airways became the first of the trio to be fully privatized. Even today, some countries maintain partial ownership of their national airlines [e.g. the French state maintains a minority share in Air France (Air France KLM 2016). Landing rights were essentially allocated in a tit-for-tat mode between the European countries and the U.S. (Cento 2009).

While this setup helped the governments to guarantee a certain level of access to air travel, it also kept competition at bay. Prices were essentially calculated using a cost plus approach, guaranteeing the companies' financial health (Pompl 2007).

In 2008, the transatlantic market of airline traffic was liberalized through the Open Skies Agreement between Europe and the U.S. In addition, the emergence of Low Cost Carriers (LCC) and consolidation of the big Full Service Carriers (FSC) led to a change in the mechanics of air traffic and unit cost became more and more important.

In addition, following the tremendous GDP and population growth in other parts of the world, the center of gravity of global air traffic moved more and more eastwards. While the non-Western world accounted for only 24 % of Revenue Passenger Kilometers (RPK) in 1970 (Airbus GMF 2011), just Asia and the Middle East alone represented 31.4 % and 8.9 % respectively in 2014 (ICAO Annual Report 2016).

All in all, air traffic has doubled every 15 years causing a corresponding rise in demand for capacity and therefore a significant increase in aircraft sales and production. Nevertheless, competition and high volatility in demand, linked to economic developments, oil price volatility and catastrophic events like 9/11, SARS and armed conflicts, have globally caused limited airline profitability. On average, Airlines only reached break-even between 1990 and 2012 and only bounced back to an average EBIT margin of 7.7 % in 2015 (Pearce 2015).

1.2 Aircraft Makers and Suppliers

In the 1960s, the world was in the midst of the cold war and especially in Europe, aerospace was first and foremost a key industry in the arms race. Therefore, many companies in this sector were focused on state-funded defense projects (Schmitt and Gollnick 2016). Even commercial aviation projects such as the Caravelle in France or the VFW 614 in Germany were at least to some degree made possible through state support. While the sales of civil aircraft were dominated by U.S. companies, even these made a large part of their business through defense projects (Wittmer et al. 2011).

In order to overcome the U.S. dominance in the civil aircraft market, France and Germany joined forces in 1969 and formed the Airbus consortium which was officially set up in 1970. While the partner-companies remained independent legal entities, they pursued a common aircraft project, the A300. The joint project was governed by work-share agreements, somewhat similar to what had already been done in trans-national defense projects such as the Transall. A small central entity, called 'Airbus Industrie' was at that point in time in charge of the overall coordination as well as sales and customer service (Airbus 2016).

Over the years, further aircraft types such as the A310, A320 and A330/A340 were added and sales climbed due to the growing market as described above but also due to the increase in market share for Airbus. In 2000, Aerospatiale of France, German DASA and Spanish CASA, merged into EADS. This new company owned 80 % of Airbus S.A.S. while British BAE SYSTEMS held the remaining 20 % (Airbus 2016). The name change to Airbus Group alongside a significant reduction of political influence in 2013 was the last step in European aircraft manufacturer consolidation as can be seen in Fig. 1.

The American manufacturers of large civil aircraft also went through a long consolidation phase. When Lockheed decided to exit the large civil aircraft market in 1981 after the commercial failure of the Tristar, only two American manufacturers remained. These two, McDonnell Douglas and Boeing finally merged in 1997 at a time where Douglas' market share had fallen below 20 % in the U.S., so that only Boeing commercial aircraft remained. Russian manufacturers could not match the competitiveness of the Western manufacturers after the end of the Cold War and disappearance of the protected communist markets (Schmitt and Gollnick 2016).

Airbus managed to secure a significant portion of the commercial aircraft market and today holds a market share of approximately 50 %, which can be seen as a true European success story. In parallel, aircraft production rates increased considerably, causing a change from job shop production to small series production. The A300, for example, reached its maximum annual output in 1982 at 46 deliveries,

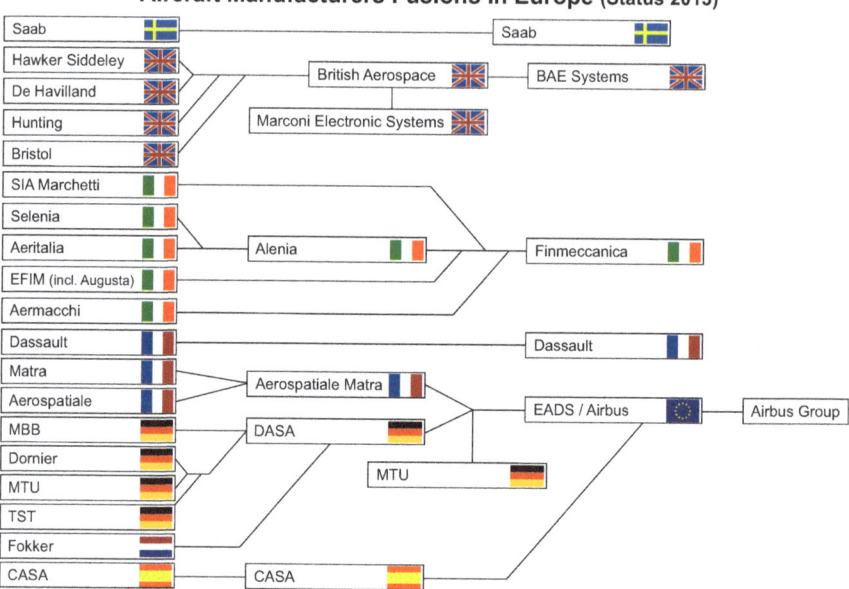

Fig. 1 Consolidation of European aircraft manufacturers (Schmitt and Gollnick 2016)

while the A320 Family today is moving to a monthly production rate of 60 aircraft by 2019. This also had significant influence on the supply chain, since it needed to cope with not only an increased number of shipsets but also a more industrialized setup. At the same time, aircraft OEMs reduced their degree of vertical integration, requiring suppliers to take over larger work packages and also to enter into risk-sharing agreements including their own design responsibility, which further supported the consolidation of the supplier base (Schmitt and Gollnick 2016).

The challenges in the airline world described in the beginning of this chapter result as well in cost pressure on aircraft OEMs and subsequently suppliers. This further fosters the transformation to a more industrialized setup that leverages scale efficiencies.

1.3 Motivation for This Book

The idea of the book is to describe the challenges for the commercial aviation supply chain along the entire product value chain in order to allow a holistic view on the drivers for transformation and the changes that have either been implemented already or will probably need to be implemented in the near future.

In order to get a complete overview, the book is composed of a set of articles written by experts and practitioners from academia, suppliers, consulting companies and Airbus.

The book is structured along five main chapters that are following the aircraft lifecycle starting from the product development phase including a detailed view of the market situation as well as technical developments and innovation. In the chapter thereafter, the book focuses on configuration and demand, highlighting the specificities of customization in the aviation industry. It continues with a chapter about component manufacturing followed by a section about assembly and integration activities and finishes with a chapter about aftermarket business models.

2 Overview of the Book

2.1 Product Development

The chapter "Product Development" deals with the evolving market environment from the customer perspective as well as with new technological developments that have a significant impact on aircraft operations.

As already outlined in the introduction, the market environment for airlines has gone through a series of notable changes and continues to evolve further. While more and more people can afford air travel and therefore induce a significant growth in RPK year-on-year (IATA: 6.5% RPK growth 2015; 10 year average 5.5% p.a.), cost pressure still increases due to the price sensitivity of the new passenger groups as well as the emergence of competition through low cost carriers resulting into lower yields per Available Seat Kilometer (ASK) (IATA: −4% in 2015).

Figure 2 shows the cost distribution of airlines based on an International Air Transport Association (IATA) special working group poll of 55 international airlines for the year 2013. The biggest single cost factor for airlines is fuel and oil, accounting for more than 33%. The next biggest block is the actual ownership of the aircraft, which is primarily composed of either depreciation charges or leasing rates. The next most important cost factor is maintenance. A closer look at these three and the other cost elements reveals that they are either independent or highly underproportional to the number of passengers on board. For fuel burn, for instance, the only marginal cost for an additional passenger on board would be the extra fuel consumed due to the additional payload. Likewise aircraft ownership is virtually independent of the number of passengers that fly on the plane, as is maintenance expenditure. Figure 3 shows the same cost breakdown as Fig. 2 but costs that are totally or at least to a very large degree independent of the number of passengers on board are highlighted in color. Altogether, they represent almost 70% of the total airline cost.

A consequence of this is the attempt to install more seats into a given aircraft in order to reduce the Cost per Available Seat Kilometer (CASK) which has become a major Key Performance Indicator (KPI) for competitiveness. This can partially be

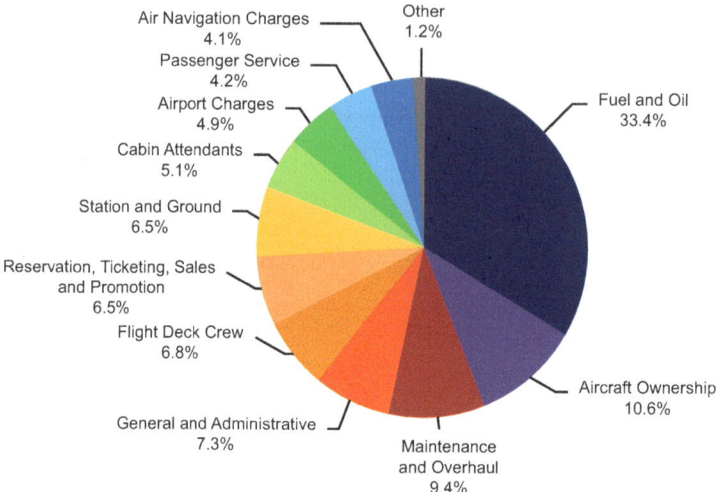

Fig. 2 Distribution of airline cost (cf IATA 2014)

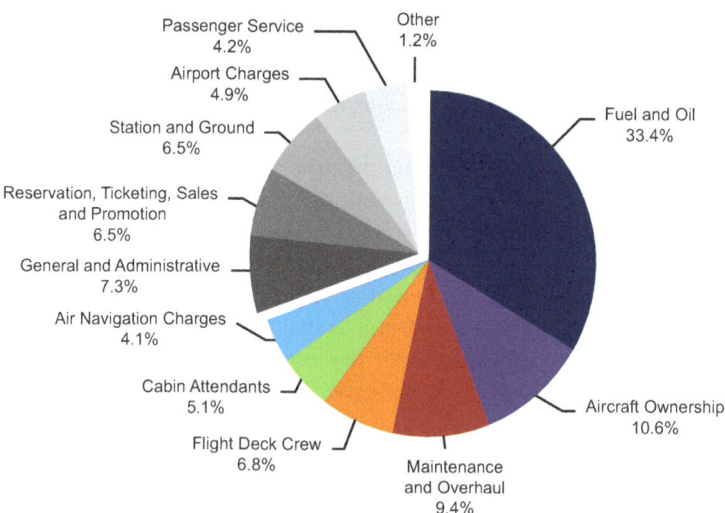

Fig. 3 Leverage of cabin densification (cf IATA 2014)

achieved through a reduction of seat pitch but, all else equal, it would also result in a reduced comfort level in the same cabin class. Over the last years, there have been some more creative approaches to this problem that combine an equivalent level of passenger comfort with additional seat count in the cabin. For instance, ultra-slim seats are today widely used in newly built aircraft and for retrofit in order to reduce the seat pitch by decreasing the thickness of the backrest. While this only frees-up some centimeters per row, over the entire cabin length it can easily result in one additional seat row. Furthermore, the positioning and integration of cabin monuments has seen significant changes with more integrated galley and lavatory solutions, underfloor concepts, as well as better usage of the space reserves in the traditional layouts, for example in front of the rear pressure bulkhead.

In order to illustrate the positive effect of improved cabin space efficiency on aircraft productivity, let us take a simplified example. For an aircraft with a seat count of 190 today, through a combination of the space efficiency measures described above, we can add one row of six seats bringing the seat count to 196. As shown in Fig. 3 and under the assumption that the changes in fuel burn etc. are negligible and that no additional crew is needed (since the threshold for additional flight attendants—1 per 50 passengers—is not hit), the cost per seat decrease significantly as around 70 % of the total airline cost can be assumed not to be impacted by additional passengers. Assuming an original cost of 100 units per seat, this would reduce to less than 97, i.e. a saving of more than 3 % per seat.

The success of the A320neo launch has shown the impact that new engine technology can have on aircraft fuel efficiency, noise, emission levels and therefore ultimately on success in the market. Further technology developments such as ultra-high-bypass engines or new engine materials are expected to further enhance the fuel burn performance of aircraft propulsion and overall aircraft operating cost (NASA 2011).

The third area, which will still contribute significantly to increased aircraft performance, is lightweight design and technology. Probably the most promising new development in this domain is Additive Layer Manufacturing, commonly known as 3D printing. Besides the ability to easily generate a physical part from digital 3D data without having to produce tooling, the main potential lies in new, bionic structures that will allow for significant weight savings through the use of hollow shapes and other geometrical layouts that cannot be manufactured today, at least not in an industrialized mode. In order to be able to fully utilize these potentials, new simulation and calculation methods will have to be developed and certified with significant impact on current design practices (Gibson et al. 2015).

2.2 Configuration and Demand

The chapter "Configuration and Demand" explains the unique importance for the aircraft industry of a customer specific configuration or customization in comparison to other industries and its implications on the supply chain. A particular emphasis is on the applicability of automotive concepts for aerospace.

In order to grasp the effects of customization on aircraft production, it is important to understand the underlying business mechanics. When an airline signs a purchase agreement for large commercial aircraft, the actual aircraft specification is far from complete. Since in most cases several years go by between the contract closure and the actual delivery of the aircraft, the airline reserves the right to modify its specification until a certain point in time before the industrial delivery of the so-called Head of Version (HoV), i.e. the first of several (more or less) identical aircraft built for a given customer.

While some of these modifications concern system or engine options, a significant proportion concern the cabin. Airlines constantly strive to optimize the cabin for the routes they are flying, the kind of customers they want to attract and the brand image that they try to convey. For all these aspects, the cabin is the major distinguishing factor. Besides these marketing oriented factors there are other aspects that play a role in the modification selection process for airlines, such as fleet commonality and maintainability.

While customer specific options for equipment and interior also exist in other industries such as passenger cars, commercial trucks and buses, the customization process for commercial aircraft goes further: not only can the customer select from a pre-defined catalogue of options, but he may actually request technical changes that require additional design work from the aircraft maker and the relevant suppliers. This means that every HoV is a unique mix of the standard aircraft configuration, selected catalogue options and modifications that have been specifically developed for this customer. To further complicate matters, the airline additionally has the choice between so called Supplier Furnished Equipment (SFE) and Buyer Furnished Equipment (BFE) for many cabin items.

In a nutshell, SFE means that the airline buys equipment, such as galleys, lavatories etc. from the aircraft manufacturer, who takes care of design and qualification and is then responsible for timely production through his supply chain. In the case of BFE, the airline buys directly from a supplier, who is then responsible for design and production and on-time delivery to the aircraft maker for embodiment. BFE gives a lot of freedom to airlines to adapt the aircraft to its needs, and it creates a complex three-party relationship that needs to be carefully managed. For the sake of completeness, it should be mentioned that there have recently been attempts to create hybrid models between BFE and SFE that aim at combining the advantages of both models.

As described above, the customization process for aircraft is complex and every aircraft, or at least every Head of Version, is a project in its own right. However, other than in the 1970s and 1980s, aircraft today are built in a small serial production mode with outputs reaching more than 500 a year for some aircraft types. In order to illustrate the importance of this achievement, let us make a simple comparison: on average, a new aircraft costs in the range of 100 MUSD. If we look for other industries that reach similar levels in terms of product value, there are few that compare. For 100 MUSD, you can also get a sizeable office building with about 30 floors.[1] While many office buildings may have similar layouts and make use of

---

[1]Assuming a standard European office building with 30 floors (each approx. 1500 sqm), the estimated costs for the building itself will be approx. 100 MUSD (2000 USD/sqm).

many standardized elements, they are still heavily customized. Similar to an aircraft, it takes more than a year from the specification freeze to the actual delivery or more accurately, inauguration, in this case. Furthermore, buildings as well as aircraft projects are subject to tight regulation and safety requirements, which add another layer of constraints. So one could argue that the industries are very similar, however, an aircraft maker is completing around three of these projects a day!

As mentioned before, customization is to a significant degree focused on the cabin. However, in order to integrate the cabin into the aircraft, there are also knock-on effects on other work packages such as electrics, water/waste and structure, including counterbalancing measures to adjust the aicrafts' center of gravity after cabin design completion. To some extend this explains the long lead-times necessary between a specification freeze and the actual cabin installation since some of the needed modifications already need to be considered very early on in the build process, not only at Airbus but also at the suppliers. The combination of small series production with a high degree of customization therefore requires a supply chain that is flexible enough to develop and produce many customer configurations at the same time, while maintaining industrial standards in terms of on-time and on-quality delivery in order not to jeopardize the production process at the aircraft OEMs.

One of the key elements to successfully manage the supply chain is a highly frontloaded demand management, i.e. to constantly ensure that the supply chain has sufficient and timely visibility of what variants it needs to deliver but also to take supply chain capacity constraints into account for production planning at the aircraft maker. Due to long lead times and sometimes very limited capacity of small specialist companies, a significant change in the product mix for certain modifications or options may actually result in delivery problems.

2.3 Component Manufacturing

The chapter about component manufacturing highlights specific aspects of interest in dedicated articles. One example illustrates the implications of globalization on the aerospace supply chain. Another article outlines the strategy of a supplier to vertically integrate several steps along the value chain from raw material provision to parts machining, which can significantly cut cost.

Globalization has accelerated enormously in the last decades and is the dominating driver in many industries, such as automotive and power generation. Emerging economies such as the BRIC states (Brazil, Russia, India and China) have benefitted from strong GDP growth as well as a growing importance of their domestic markets fueled by a large population, a growing middle class and to some extend the profits from natural resources such as fossil fuels, metals and minerals. At the same time, some countries especially China and India, have maintained a cost competitive manufacturing base with low labor cost and growing industrial capabilities (Dicken 2015).

In addition, many emerging countries require Western companies to invest in so called "industrial cooperation" in return for the goods and services that they are exporting to them. This often results in incentives or obligations to build up local suppliers for strategic industries such as automotive or aerospace.

For the established Western players in the long run this means additional competition. However, so far the growth generated by the globalization of the market has outweighed competitive redistribution significantly. Looking at the various aerospace commodities, we observe significant differences in terms of penetration of non-Western companies: for Aerostructures and Detail parts, companies from emerging countries such as AVIC (China), TAI (Turkey) and Dynamatics (India) have managed to become tier-1 suppliers or small-airframe manufacturers. In the Equipment, Systems and Cabin area, many production and assembly facilities have been opened in emerging countries but those are often subsidiaries of the established Western players. Up to know, there are very few examples of companies from emerging countries that act as tier-1 or system integrator in this area, e.g. Larsen & Toubro or TAI. On the aircraft OEM level, a company like Embraer, which has successfully entered the market for regional jets, can be seen as remarkable exception. In the segment of commercial jets above 150 seats, only China's AVIC has launched a project, the COMAC 919, which is planned to enter into service in the next few years (Eriksson and Steenhuis 2016).

2.4 Assembly and Integration

The chapter "Assembly and Integration" deals with current concepts that enable a smooth serial production flow at the aircraft OEM. More specifically, it discusses the importance of Quality Gates between the various steps of the value chain, starting at design deliverables and continuing onwards throughout production and final assembly. Furthermore, it elaborates on the principles of lean production as well as the upcoming opportunities and changes arising from the so called Industry 4.0 or digitalization in manufacturing.

In theory, deliverables throughout the commercial aircraft value chain are handed over to the subsequent step 100% on-time and on-quality in order to ensure a smooth process. However, given the complexity of an aircraft and the fact that some sub-assemblies or pieces of equipment may already constitute values of several million, the concept of an either good or bad component and quality measurement by parts per million (ppm) has proven to be too rigid. So rather than using a simple binary logic and rejecting every work package that does not fully meet the specification, a concept of so called Quality Gates has been introduced at Airbus.

A Quality Gate defines allowable thresholds for the acceptance of a deliverable from further upstream in the chain, be it internally within Airbus or from suppliers. As mentioned earlier, this concept is not only applied to physical parts but also to engineering deliverables. The supplier (internal or external) and the customer jointly define limits within which a deliverable can still be accepted without jeopardizing the process further downstream. For a section delivered to the Final Assembly Line (FAL), this can, for instance, be the amount of hours of outstanding work that is required to be completed during the FAL process for tasks that should have been completed at the major component level assembly.

Other elements that have become increasingly important are lean production concepts. With the higher production rates, many proven lean manufacturing concepts from other industries could be applied to aerospace. While these elements need to be adapted to the specifics of the aircraft industry, the basic principles remain the same.

Especially the automotive industry is generally seen as one of the industries with the most advanced supply chain concepts. Some of these, such as kitting for specific assembly takts or Kanban, have in the past been successfully applied in aeronautics as well. However, due to significant customer benefit achieved through highly customized configurations, many concepts cannot simply be copied but either need to be adapted or cannot be used at all. While component pre-assembly stations for fully flexible, last-minute customization to follow the main assembly sequence can boost productivity in an automotive environment, the concept cannot be applied easily to jet manufacturing, where customized configurations are often specifically designed and have to be validated by engineering.

The next mega-trend in production will be changes driven by the potential of digitalization and highly sophisticated automation throughout the value chain. DMU concepts (digital mock-up) for the geometrical integration of an aircraft are already commonly used today during the development phase. In the future, increasingly refined digital aircraft models will also allow functional simulation and testing of an entire aircraft or significant sub-systems in a "virtual aircraft" model. Modelling interdependencies between the various subsystems and the combined simulation of work packages that are designed by different suppliers will help shorten development cycles. In addition, design-to-automation concepts will be a key differentiator in production cost. Robotics, automation and augmented reality enhanced design and assembly concepts are foreseen to reduce the cost of production significantly (Erbe n.d.).

2.5 *Life Cycle Business Models and Aftermarket*

The chapter "Life Cycle Business Models and Aftermarket" deals with the in-service life of the aircraft following its delivery. A focus is put on the difference between the propulsion and the aircraft OEM business model over the lifecycle and opportunities for collaboration between the two. In addition, it discusses the spare parts market and the importance of material services and aftermarket integrators. Furthermore, the role of predictive maintenance concepts and the resulting opportunities are addressed.

For an aircraft OEM, the vast majority of the cash flow it generates with an aircraft throughout its lifecycle is generated at aircraft delivery. So called pre-delivery payments (PDPs) at contract closure and at certain milestones during the build process and the final payment at delivery constitute a very high percentage of the total revenues an aircraft maker generates with an aircraft. In the traditional model, the aftermarket income of a civil aircraft manufacturer is very small compared to the business of selling aircraft. Aircraft makers have started to enter the aftermarket business through Power-by-the-Hour services such as the Airbus Flight Hour Services, or the acquisition of specialized aftermarket companies. However, the weight of these new activities is still marginal compared to the fast growing overall OEM business (Industry & Analysis 2015).

Engine makers have a business model which is somewhat inverse: when selling their engines, they try to sell a maintenance package for their product, too. They are ready to compromise on the revenue generated with the actual engine sale due to the significant cash flows that they can expect from the in-service activities that they perform themselves or through a partner network.

The different business models between aircraft OEMs and engine makers offer interesting opportunities for collaboration between the two in terms of revenue sharing at different stages of the lifecycle in order to smoothen the cash flow for both of them and therefore reduce the need for up-front financing for the engine makers and allow the aircraft maker to benefit more from the aftermarket revenue stream of the platforms (Rhoades 2014).

As for the engines, the aftermarket also plays a crucial role for many aircraft systems and cabin equipment. From an airline standpoint, having the right spare parts at hand whenever needed is a key factor of success, since a so-called AOG (Aircraft On Ground) i.e. an aircraft unable to fly, caused by unavailability of spares, means that revenue is lost and passengers may need to be compensated.

Besides suppliers managing their aftermarket business themselves, distributors such as Aviall and Satair have successfully established themselves in this market as agents between the manufacturer of a given part and the airline. They offer value adding services to the suppliers on one hand and to the end user, typically airlines or MROs (Maintenance, Repair and Overhaul) on the other. They specialize in the aftermarket business, which is significantly different from serial production.

In serial production demand is normally deterministic, meaning with some exceptions, it can be derived from the customers' production schedule. Additionally, there is only one or at least very few customers, while in the aftermarket spare part needs come from hundreds or thousands of different customers in multiple locations. Prediction of spare parts needs can only be done on a stochastic basis and to make things even more challenging, spare parts may also be required for aircraft that are not in production anymore (Emmanoulidis et al. 2013).

Distributors play a crucial role in this market: they maintain a stock of parts and therefore allow the airlines or MROs to reduce their own stocking levels. At the same time, their planning models help predict the demand, and by buffering the volatility of the aftermarket, they help the suppliers to maintain a steady production flow. Furthermore, they normally offer around the clock services alongside high performance logistics processes that are suitable for aftermarket needs. Since they typically take care of a wide spectrum of parts, they also offer significant scale efficiencies compared to a small or medium size supplier that is managing its own spare parts operations (Chakravarty 2014).

As mentioned above, the low predictability of maintenance and the need for spare parts make the aftermarket somewhat volatile and leads to overall inefficiencies in the supply chain and consequently also in aircraft operations. Non-availability of aircraft due to in-service issues, high buffer stocks throughout the industry (worth approx. 50 BUSD, Arbor 2010) and long lead-times for spareparts are just a few examples of these issues (Fransoo et al. 2011).

Over recent years, engine makers, aircraft OEMs, airlines and MROs have started to make use of data that is obtained from in-service aircraft to improve the visibility of future maintenance and spare parts needs. These schemes are often referred to as predictive maintenance and are intended to reduce maintenance costs and increase overall asset utilization (Capgemini 2009).

The rise of so-called big data applications, i.e. the analysis of huge amounts of structured and unstructured data in order to find repetitive patterns, further complements these schemes. Better forecasting of both maintenance events and spare parts needs has the potential to significantly lower costs and increase aircraft reliability.

2.6 Outlook

The chapter "Outlook" deals with the long-term development of the aviation industry. A particular emphasis of the chapter, contributed by the BDLI, (German Aerospace Industries Association) deals with the chanes, which will occur in the supply chain due to technical innovation and global developments.

Several possible scenarios for the industry are outlined including some extreme developments. However, the majority of experts assume a "standard" outlook, meaning a continued long-term increase in the number of passengers and flight revenue kilometers. The economic growth including an emergence of the middle class, especially in Asia, will lead to a rising demand in the entire commercial aviation industry.

In the long term, airlines will require steady improvement of the product to remain attractive and competitive in the market. For aircraft OEMs this includes the necessity to continue developing technologies to reduce CO_2 and noise emissions. Another challenge will be to provide the most innovative cabin with visionary in-flight concepts to enable airlines to differentiate from their competitors.

The structure of the commercial aviation industry will change. The supply chain itself will gain more flexibility, such as less rigid and stratified structures. For medium-sized suppliers this could lead to a loss of competitiveness and a downgrade in the supply chain due to their small role in international programs.

However, suppliers investing in growth, globalization and digitalization will be strengthened.

Ongoing demand to deliver highest quality on time with increasing cost pressure and more complex requirements from customers' side will certainly continue. Concerning technology and innovation the aerospace industry continues to be seen as a pioneering model for other branches.

Concluding, we would like to thank all experts for contributing to this book with their long-standing experience. We hope to give you many interesting insights into the future potential and limits of supply chain integration in the commercial aerospace industry. Enjoy reading!

References

Air France KLM (2016) Shareholding structure. Available via http://www.airfranceklm.com/en/finance/financial-information/capital-structure. Accessed 09 Mar 2016

Airbus (2011) Airbus General Market Forecast (GMF) 2011–2030. Available via https://www.airbusgroup.com/dam/assets/airbusgroup/int/en/investor-relations/documents/2011/Presentations/2011-2030_Airbus_full_book_delivering_the_future.pdf. Accessed 26 Feb 2016

Airbus (2016) The narrative. Available via http://www.airbus.com/company/history/the-narrative/. Accessed 26 Feb 2016

Arbor A (2010) 2009 commercial air transport MRO logistics survey. In: OEM feedback report. Aero Strategy Management Consulting. Available via https://www.iata.org/whatwedo/workgroups/Documents/mro-logistics-survey-report-v3_2010.02.22.pdf. Accessed 23 Mar 2016

Capgemini (2009) Maintenance Repair and Overhaul (MRO). Aerospace and Defense. Available via https://www.it.capgemini.com/resource-file-access/resource/pdf/Maintenance__Repair_and_Overhaul__MRO_.pdf. Accessed 16 Mar 2016

Cento A (2009) The airline industry: challenges in the 21st century. Physica-Verlag, Heidelberg

Chakravarty AK (2014) Supply chain transformation. Springer, Berlin

Dicken P (2015) Global shift: mapping the changing contours of the world economy, 7th edn. Sage, London

Emmanoulidis C, Taisch M, Kiritsis D (2013) Advances in production management systems. Springer, Heidelberg

Erbe H (n.d.) Technologies for cost effective automation in manufacturing (low cost automation). In: IFAC professional brief. IFAC, Center for Human-Machine Systems Technische Universität Berlin. Available via http://www.ifac-control.org/publications/list-of-professional-briefs/pb_erbe_final.pdf. Accessed 9 Mar 2016

Eriksson S, Steenhuis H-J (2016) The global commercial aviation industry. Routledge, New York

Fransoo JC, Wäfler T, Wilson J (2011) Behavioral operations in planning and scheduling. Springer, Berlin

Gibson I, Rosen D, Stucker B (2015) Additive manufacturing technologies. Springer, New York

IATA (2014) Airline cost management group report FY 2013. Available via http://www.iata.org/. Accessed 2 March 2016

ICAO (2016) Annual report 2014. Available via http://www.icao.int/annual-report-2014/Pages/default.aspx. Accessed 23 Feb 2016

Industry & Analysis (2015) Aircraft parts. In: 2015 top markets report. International Trade Administration. Available via http://trade.gov/topmarkets/pdf/Aircraft_Parts_Top_Markets_Report.pdf. Accessed 9 Mar 2016

NASA (2011) The promise and challenges of ultra high bypass ratio engine technology and integration. In: Environmentally responsible aviation project integrated systems research program. National Aeronautics and Space Administration. Available via http://www.aeronautics.nasa.gov/pdf/asm_presentations_promise_and_challenges1.pdf. Accessed 14 Mar 2016

Pearce B (2015) 2015 end-year report. IATA. Available via http://www.iata.org/economics. Accessed 15 Feb 2016

Pompl W (2007) Luftverkehr, 5th edn. Springer, Berlin

Rhoades DL (2014) Evolution of international aviation. Ashgate Publishing, Farnham

Schmitt D, Gollnick V (2016) Air transport system. Springer, Wien

Wittmer A, Bieger T, Müller R (2011) Aviation systems. Springer, Berlin

Klaus Richter is Chief Procurement Officer for Airbus Group & Airbus. Additionally, he serves as Chairman of the Board of Airbus in Germany and leads the supervisory board of Premium AEROTEC Group. Richter's professional path within Airbus Group dates back to 2007 when he joined Airbus S.A.S. as Executive Vice President Procurement.

Before joining Airbus Group, he was Senior Vice President Direct Materials Purchasing for BMW.

He started his professional career within McKinsey & Company where he worked as management consultant for automotive, electronics & aerospace businesses and product development.

He graduated from the Technical University of Munich, Germany, with a doctorate in mechanical engineering and spent 2 years as a researcher and teacher at the University of California at Berkeley.

Born in Munich in 1964, Richter is married with two children and currently lives in Toulouse, France.

Nils Witt is Head of Business Operations & Strategy within Airbus Procurement Operations. Prior to this, he was Executive Assistant to Klaus Richter, Chief Procurement Officer Airbus Group & Airbus. Before joining Procurement, Nils worked in Airbus Customer Service, where he was leading a department in charge of developing and selling aftermarket supply chain solutions. Previously, he worked in Services Strategy & Business Development focussing on services profitability improvement and M&A cases for the aftermarket supply chain sector. He originally started his career in Airbus as an Industrial Planning Engineer in the A380 Programme.

Nils holds a Master of Science degree in Industrial & Systems Engineering from the University of Florida. Born in Kiel, Germany, in 1982, Nils today lives and works in Toulouse, France.

Part I
Product Development

Aviation's Future Is as Bright as Its Past

Andrew Gordon

Abstract The need to travel is one of mankind's most enduring traits, but it has only been in the last 100 years that flight has enabled him to traverse great distances in very little time. Add to this the fact that aviation is increasingly affordable and you have all the ingredients for an industry that connects economies, countries, cities and more importantly people. Its demand is evident in the industries statistics, with air traffic historically doubling every 15 years. With growing GDP, increasing wealth, greater liberalization, and more capable eco-efficient aircraft, air traffic is forecast to double again in the next 15 years, with the world's aircraft fleet also forecast to double over the next 20 years. Aviation will continue to contribute to nations' economies delivering both GDP and jobs, directly and indirectly, through tourism for example, where half of all tourists use aviation to get to their holiday destinations. In the future, new technologies will simplify and enhance the passenger experience, and will offer manufacturers opportunities to improve their products as well as their design and manufacturing capabilities.

1 Putting Aviation's Place in Transportation History into Perspective

At its most basic level aviation is a means of transportation, a method by which people are able to make journeys to meet their individual needs. Since the dawn of man, people have had the need to travel. In ancient times this need was primarily in search of food and security, then for trade and in search of wealth. More recently we also travel for leisure, to experience different cultures or simply to escape to climates more suited for relaxation and recreation than our own. Man's early journeys were on foot, this as long ago as Cro-Magnon man in 30,000 BC, who at that time were flourishing and beginning a slow migration from the near east into Europe. The option of riding a horse only became possible with their domestication

---

A. Gordon (✉)
Airbus, Toulouse, France
e-mail: andrew.gordon@airbus.com

Fig. 1 The A350 cabin, a long way from the first passenger experience. Source: Airbus, Master films, P. Masclet

at around 4000 BC. Journeys improved further with the invention of the wheel around 3500 BC, and sailing ships in 3100 BC. For the next 5000 years, people were limited to journeys which were slow and often made difficult by geography and the weather. How many times must a traveller have looked at the birds in the sky and wished they were able to join them to traverse the mountain ranges, seas and vast plains before them? This apparently impossible dream finally became a reality just over 100 years ago, with the first paying air passenger taking a flight across Tampa Bay in Florida in January 1914. This airboat service, between St Petersburg, USA and Tampa, took 23 min and cost $5 (one-way), or the equivalent of about $116 in today's dollars. The same journey by boat took 2 h, by car 20 h and 4 h by train. The benefit of aviation was clear and people were prepared to pay for it.

Over the next 100 years, aviation has grown from this first passenger to some 3.5 billion passengers in a single year in 2015, with aircraft departures reaching 34 million globally (ICAO 2015). The network and service has gradually grown with a series of firsts. The first permanent airport and commercial terminal used solely for commercial flights opened in 1922, the first North American transcontinental flight in 1923, the first commercial flights across the Pacific in 1936 and the Atlantic in 1939. In 1952, the De Havilland Comet became the world's first commercial jet airliner, Concorde flew its first supersonic passenger flight in 1976, the Airbus A380, the first full double deck aircraft, entered commercial service in 2007. The industry has come a long way in a very short time. In the 32,000 years man has journeyed, he has only been able to use the sky, the third dimension, for a mere fraction (Fig. 1).

2 Market Drivers, a Global Context

Air travel has grown dramatically, with aviation traffic measured in Revenue Passenger Kilometres (RPKs), that is the number of revenue paying passengers multiplied by the distance that they travel, doubling every 15 years, despite a number perturbations caused by economic slowdowns or other shocks. Each time one of these events occurred, air traffic has bounced back, revealing the value that passengers place on air travel. Despite industry downturns coinciding with the 9/11 attacks, the Severe Acute Respiratory Sydrome (SARs) and the financial crisis in 2008/2009, aviation has still managed to grow 98 % (Airbus 2015) by the end of 2015. What is behind this impressive growth? Economic growth is often cited as a key driver for aviation growth with the relationship in this decade being a 1 % year on year increase in global GDP translating to 2 % in air traffic growth. Growth in population is another factor which is often used in traffic forecasts as a descriptive variable for air traffic growth. A third variable is airline yield, which is the amount of revenue airlines earn as a proportion of the traffic they carry. Typically if yields decline, normally through fare reductions, this also has the effect of stimulating air traffic growth. But beyond these measures and datasets used by forecaster's, fundamental changes to the world in which we live are also having the effect of driving the demand for aviation both today and in the future.

One such factor is increasing urbanization. Globally, more people live in urban areas than in rural, with 54 % of the world's population residing in urban areas in 2014. In 1950, only 30 % of the world's population was urban, by 2050, 66 % of the world's population is projected to be living in cities according to the United Nations (United Nations Population Division). This fundamental shift in people from rural to urban locations is being driven by a desire for improved quality of life coming from better job opportunities, education, and living standards. These in turn help to drive wealth. One indicator is the projected number of middle class people, particularly in emerging/developing nations. Over the next 20 years the number of people who can be classified as middle class is expected to double in these nations (Kharas 2012; Airbus 2015) (Figs. 2 and 3).

This increased wealth is expected to translate into increased disposable income and level of personal consumption. All things helping to drive demand for aviation as it becomes increasingly economically accessible to a greater share of the world's population. It is therefore unsurprising that stronger air traffic growth is forecast in these countries and their regions than in more mature markets. With their 6.2 billion people, the emerging markets are expected to witness +5.8 % air passenger traffic growth over the next 20 years (Airbus 2015). The more advanced nations like those in Europe and North America with their one billion people will be growing by 3.8 % over the same period. As wealth increases so does the propensity of people to fly.

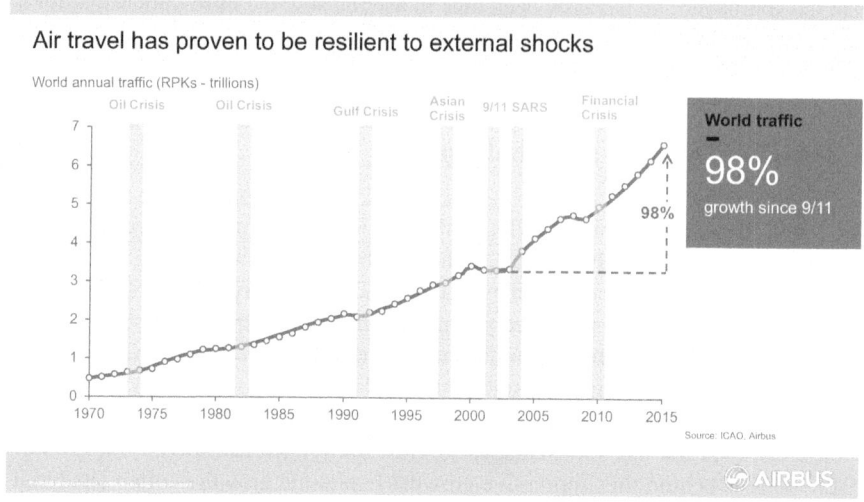

Fig. 2 The need to fly means that commercial aviation is resilient. Source: Airbus GMF (2015)

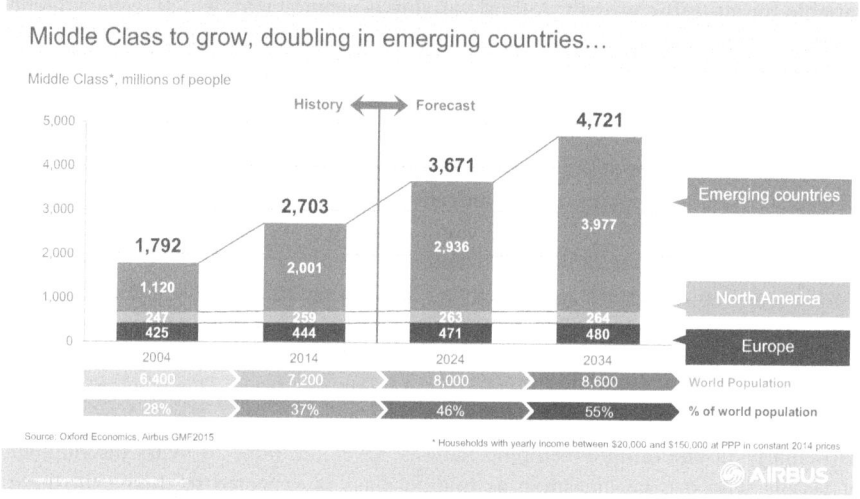

Fig. 3 A growing global population and wealth will help drive aviation's growth. Source: Airbus GMF (2015)

Today, 25 % of the population from emerging countries take a flight every year, in 2034, this number will increase to nearly 75 %, with China's very large population to have roughly the same propensity to travel by air as Europeans today (Airbus 2015).

As well as the accessibility of aviation increasing through increasing wealth, aviation has been able to grow through the positive actions of governments who also see the value aviation can bring. An important example of this is a focus on the deregulation and greater liberalization of markets. This can be achieved in various ways, for example, through countries coming together to agree regional open skies agreements as, for example, in Europe and between the ASEAN countries, today. Individual countries can sign bi-lateral agreements with others to give greater access to each other's skies; a process championed by the US in the 1990s with new agreements signed almost every year, with a further four signed in 2015. In recent years, China has been signing agreements across the globe, agreements which in part will help to facilitate its 'Belt and Road' (english.gov.cn 2015) economic initiative.

As well as air service agreements, immigration procedures have been simplified between many countries, with particular focus on fostering greater business and tourism links. China, for example, has recently agreed significant accords with the US and Australia which ease the visa process for travellers in either direction, agreements which will no doubt stimulate further air traffic between them in the coming years. In fact, the number of Chinese visitors to Australia rose 21.6 % to just over a million in the 12 months ended November 30th 2015, and has more than doubled over the past 5 years (Australian Bureau of Statistics 2016). In the future, forecasts indicate Chinese tourism to Australia could be worth up to $13 billion by 2020, a revised figure as previous estimates have already been exceeded (The Sydney Morning Herald 2016).

Forecasters have long recognised the link between airline yield (revenue divided by traffic, typically measured in cents per RPK) and traffic growth. With falling yields and fares comes increased air traffic growth. In 1939, for example, a round trip from New York to France would have cost about $6000; today the same would cost about $500 (Askapilot.com and Airbus 2013). Globally, the average return fare (before surcharges and tax) of $375 in 2016 is forecast to be 61 % lower than 21 years earlier, after adjusting for inflation (IATA 2015).

Increasingly open skies have meant that new airline business models have been able to develop and offer more options for the world's air passengers. The most visible are the so called low fare or low cost airlines (LCCs), who have been one of the drivers of lowering average airline yields, but at the same time helping drive passenger traffic growth. These airlines, with their focus on operating the most cost efficient types, maximizing utilization and unbundling their product offering were able to start in the world's largest single domestic market, due to liberalization in the US. They spread further with European deregulation, and have gradually expanded the scale and scope of their operations with long haul operations now very much a focus of a number of low cost operators. These operations have become increasingly viable with the latest group of long haul capable aircraft and the growth of internet access for passengers to plan and book their travel.

Airbus have forecast that through drivers such as these, traffic will double again in the next 20 years, and is expected to grow at an average annual rate of 4.6 % over the next 20 years. Airlines in the Asia Pacific region with their dynamic economies and growing populations are expected to take the largest share of global traffic with 36 %, growing from 29 % in 2014. This growth will be helped by traffic flows such

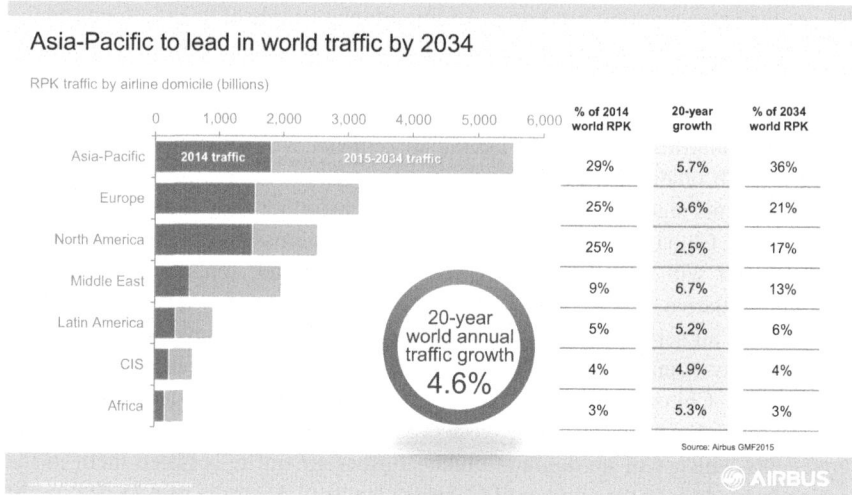

Fig. 4 Asia Pacific's airlines will carry more than a third of aviation's traffic. Source: Airbus GMF (2015)

as the Chinese domestic market which will grow to become the largest in the coming years, and domestic India which will not be the largest but will grow the most strongly. Airline traffic in Europe and North America will also grow. Despite the fact that their combined share of global traffic will fall as Asia Pacific grows, they will still carry nearly 40 % of the world's aviation traffic (Fig. 4).

What sort of aircraft will people be flying with in coming years? By far the greatest number of aircraft to be delivered will be the single-aisle aircraft carriers, which are typically equipped with 125–210 seats. If you have ever flown with a low cost carrier, or made intra-regional hops between cities anywhere in the world, the chances are high you would have been on one of these aircraft types like the A320 and Boeing's 737. Today, these types are being enhanced with the very latest technology from wings to engines, with new aerodynamic features like the sharklets on the A320neo family and the latest power plants from Pratt & Whitney and CFM International. As a result, the Airbus A320neo will improve fuel efficiency by 20 % per seat, will hold 2 tonnes of additional payload, will extend the range by up to 500 nautical miles, will lower operating costs, along with reductions in engine noise and emissions. Whilst described as an 'incremental' development, there is nothing small about these steps. Over the next 20 years, nearly 23,000 single aisle aircraft from all manufacturers will be needed. The twin-aisle market will need over 8000 new aircraft over the same period. This demand is driven by the need to replace aircraft that will retire and those needed because of industry growth, particularly for medium to long-haul operations. Some countries and regions will also increasingly need aircraft types, which range from 250 to 400 seats, to fly domestic and shorter range missions to meet demand between major conurbations in both their own countries or large cities in neighbouring countries. The Airbus A330, for example, is used in this role, and like the A320 has a NEO variant in development. The A330neo incorporates latest generation Rolls-Royce Trent 7000 engines, along with aerodynamic improvements—including new composite sharklet wingtip devices that provide 3.7 m of increased wingspan—as well as increased lift and reduced drag. It delivers fuel savings of 14% per seat compared to in-production A330s, while also providing even quieter operation, a range increase of approximately 400 nautical miles, additional payload capability, decreased maintenance costs and superior passenger comfort. Aircraft in seat categories over 400 seats like the A380, typically although not exclusively, operate on dense routes and are ideally suited for aviation routes between mega-cities. Anybody who has been to the Tom Bradley terminal at Los Angeles International Airport (LAX) won't have failed to notice the A380s lined up ready to take people to all corners of the globe. As well as being an efficient form of mass transportation typically carrying more than 500 passengers, they offer airlines the opportunity to segment their product offerings in many ways. The service comfort and quality the A380 offers passengers was highlighted when it was awarded 'Best Aircraft Type' in 2015s US Global Travel Awards. As markets grow and demand on key routes develops, the need for this aircraft type will also grow with over 1000 aircraft of this size needed in the coming 20 years.

Todays' aircraft have come a very long way from those that flew the first airline passengers just over 100 years ago. The A380, for example, is 600 times bigger, can carry 800 times more passengers, flies 80 times further than the Benoist single passenger aircraft used for this first commercial service in the US. There has been a trend for larger aircraft in the last 30 years with nearly 50% more seats offered globally, and a decrease of fuel used per 100 revenue passenger kilometres by a third in 15 years. Our industry has made significant advances over the last 100 years. For those involved today or those who will join the aviation industry in the coming years it will be their challenge to match or exceed the huge strides taken in the past, in the next 100 years (Fig. 5).

3 Benefits Beyond People

Whilst it is clear what air travel brings to the individual traveller, it is often less obvious what broader benefits aviation brings. Various industry reports help to highlight these. ATAG in their publication "Aviation, Benefits Beyond Borders" stated, for example, that the industry supports nearly 60 million jobs, contributes $2.4 trillion to the global economy, that's 3.4% of global GDP. If aviation was a country it would rank 21st in size in terms of GDP, bigger than some G20 countries (ATAG 2014). Other industry segments also benefit from aviation, for example, tourism forming an important element of many nations economic output. More than a billion international tourists are recorded travelling in 2014, more than 50% of them were carried by plane. Statistics show that international tourism (travel and

Fig. 5 The old and the new. Source: Courtesy of Gregor Kaluza for Rimowa

passenger transport) represents 30% of the world's service exports and 6% of goods and services exports (UNWTO 2015). Tourism as a worldwide export category ranks fourth after fuels, chemicals and food, and ranks first in many developing countries. Much of this benefit would not be realised without aviation today, but more importantly nor would it be in the future.

4 Things Passengers and Manufacturers Can Look Forward to

The flying experience starting with the decision to fly to returning home has significantly changed. The changes have been rapid enough for nearly everybody who has flown to be able to identify real differences. The internet's development and access, for example, has enabled airlines to develop websites and apps as an additional cost effective distribution channel. Passengers have much more control of their journey rather than relying on travel agencies or difficult phone calls with distant sales teams. Today, passengers can search for best flight and price options. The aircraft type, once invisible to the passenger, is now part of the decision making process for a growing number of passengers. Web services even advise on the aircraft layout and seats to avoid those with no windows or perhaps being a little too close to the lavatory. When booking a flight, cars, hotels, even insurance can often be acquired with a few more clicks, too. Meal choices or upgrades can be made, services like first to board can be added, seats with more legroom can be picked, more baggage added—the list of service choices will surely grow in the future. On board the aircraft, inflight entertainment systems have improved. Systems have become smaller and more powerful, boxes under the seats have been removed giving the passenger more room, as well as better quality and more choices are provided. With the proliferation of personal electronic devices, however, airlines are now be able to provide information directly to these devices either during or in advance of the flight. Technology and services in this area will continue to develop; Li-Fi a connectivity solution like Wi-Fi using light to transmit data rather than radio waves is being tested for aircraft application. It offers wide bandwidth, higher security and the use of light is particularly suited to aircraft because aircraft systems must be considered with the use of radio waves.

The additional connectivity will also benefit the crew. AirbusOn, for example, provides smart solutions such as Cabin Crew Applications (e.g. eReporting, eDocumentation, Passenger Service and Crew Interaction), cabin health and augmented/virtual reality applications to ensure seamless, paperless and connected efficiency by tablet, phone, watch or glasses on-board and on-ground in real-time.

In the future, pictures showing the destination, a city skyline or a tropical forest, could be holographically projected onto the cabins surface. A private cabin can project your bedroom at home, imitate a business conference or even a Zen garden, thanks to the projection of virtual decors. There are speculations that today's seating classes (first, economy, business) might be replaced by zones, where passengers will pay based on their desired flying experience, for example, if they want to relax, work or to be entertained during their flight.

Future aircraft could be built using bionic structures that mimic the bone structure of birds. Bone structure is both light and strong because its porous interior carries loads only where necessary leaving space elsewhere. By using bionic structures, the fuselage has the strength it needs, but can also make the most of extra space where required. This not only reduces the aircraft's weight and fuel burn, but also makes it possible to add features like oversized doors for easier boarding and panoramic windows. This is not just a dream, Airbus is already designing and building bionic structures. Today, our industry is on the cusp of a step-change in weight reduction and efficiency—with the potential to produce aircraft parts which weigh 30–55 % less, while reducing raw material used by 90 %. This game changing technology also decreases total energy used in production by up to 90 % compared to traditional methods (Fig. 6).

The future cabin's electrical systems could be compared to the human brain, with a network of intelligence pulsating through the cabin. This network will be absorbed into the structural materials, making the hundreds of kilometres of cables and wires found in today's aircraft a thing of the past. Known as 'smart' materials they can perform numerous functions, for example, recognising the passenger, so that you too are 'connected' to the plane. Materials in the cabin could well be self-cleaning, think of lotus plant leaves, where water runs off in beads, taking contaminants with it. Today, coatings inspired by this are used on the surfaces of cabin bathrooms. In the future, they will be found on seat and carpet fabrics. These intelligent materials could also be self-repairing, an application already used in

Fig. 6 The future by Airbus aircraft cabin. Source: Airbus

surface protection today. Certain paints can seal a scratch by themselves, just as the human skin can.

Some of the elements in the cabin could be created using additive layer manufacturing. The process repeatedly prints very thin layers of material until the layers form a solid object in materials ranging from high-grade titanium alloys to glass and concrete. As well as making it simpler to produce very complex shapes, this form of production wastes a lot less material than cutting shapes out of bigger blocks. While this technique is already being tested for small aircraft parts today, its use could very well be widespread in the future.

Innovations such as these will help drive aviation's development over the coming years, not just to heighten the passenger experience but to meet the challenges aviation faces today. One such challenge will be to continue to reduce aviation's impact on the environment. Today, aviation contributes about 2 % of manmade emissions, this despite the fact that aircraft produce 70 % less CO_2 and 75 % less noise than 40 years ago. It is aviation's growth that creates the challenge, a challenge that our industry has vowed to meet through commitment to tough targets. Europe's 'Flightpath 2050' calls for CO_2 emissions to be reduced by 75 %, wants to reduce emissions by 90 % and perceived noise by 65 %, all compared to levels in 2000. Airbus is already contributing to these efforts through new innovative products, with their significant reductions in fuel burn and therefore CO_2 reduction, in fact we spend over $2 billion a year to enhance aircraft efficiency.

All industry stakeholders working together will be the key to further innovations. It is useful to remember that when it comes to innovation, Jules Verne got it right:

Anything that one man can imagine other men can make real. (Verne 1873)

References

Airbus (2015) Global market forecast—flying by numbers 2015–2034. Available via www.airbus.com. Accessed 31 Dec 2015

Askapilot.com (2013) Homepage. Available via www.AskaPilot.com. Accessed 23 Jun 2016

ATAG Air Transport Action Group (2014) Benefits beyond borders. Available via www.aviationbenefits.org/. Accessed 23 Jun 2016

Australian Bureau of Statistics (2016) Overseas arrivals and departures, Dec 2015 Australia. Available via http://www.abs.gov.au/AUSSTATS/abs@.nsf/DetailsPage/3401.0Dec%202015?OpenDocument. Accessed 12 Feb 2016

english.gov.cn (2015) Action plan on the belt and road initiative. Available via http:// http://english.gov.cn/archive/publications/2015/03/30/content_281475080249035.htm. Accessed 23 Jun 2016

Freed J (2016) Annual Chinese visitor numbers exceed 1 million for first time. The Sydney Morning Herald. Available via www.smh.com.au/business/the-economy/annual-chinese-visitor-numbers-exceed-1-million-for-first-time-20160111-gm3shn.html#ixzz490HpRcEo. Accessed 13 Jan 2016

IATA (2015) economic briefing note December 2015. Available via http://www.iata.org/whatwedo/Documents/economics/IATA-Economic-Performance-of-the-Industry-end-year-2015-report.pdf. Accessed 10 Dec 2015

ICAO (2015) Provisional numbers Dec 2015. Available via http://www.icao.int/Newsroom/Pages/Continuing-Traffic-Growth-and-Record-Airline-Profits-Highlight-2015-Air-Transport-Results.aspx

Kharas H (2012) Oxford economics. Brookings Press, Oxford

United Nations World Tourism Organisation (UNWTO) (2015) Exports from international tourism rise to US$ 1.5 trillion in 2014. Available via www.media.unwto.org/press-release/2015-04-15/exports-international-tourism-rise-us-15-trillion-2014. Accessed 23 Jun 2016

Verne J (1873) Around the world in 80 days. Routledge, Paris

Andrew Gordon moved to Airbus' headquarters in Toulouse in 2001. Today, he is Director Strategic Marketing and Analysis. As part of this role, Mr. Gordon is responsible for facilitating the production of the Airbus Global Market Forecast, which details future demand for all civil aircraft of 30 seats or more. Prior to this appointment, he had worked for British Aerospace Airbus as part of the Forecasting & Market Evaluation Team, where he was involved in providing analysis for many of the market and product-related decisions. In 2001, at the creation of Airbus as a single entity, Mr. Gordon moved to Toulouse, to work directly for Airbus in the Market Forecasting & Research Group, now part of the Strategy and Co-operation team. Mr. Gordon began his career with BAE/Airbus in 1991 and prior to that, he had worked for one of the United Kingdom's largest banks. He holds a Master's degree in Marketing from the University of the West of England.

Cabin Densification: SpaceFlex2 and Beyond

Scott Savian

Abstract Airlines in today's ultracompetitive industry continuously look to reduce costs while also squeezing every possible bit of revenue out of their most valuable assets, the aircraft. Revenue generation is increased by (1) increasing the percentage of seats sold (load factor), (2) increasing the "yield" by creating seats and services that passengers are willing to pay more for, (3) operating as many flights per day as possible within the given route structure, (4) and by physically adding more seats to the aircraft. In efforts to improve load factor and yield, airlines work to create a "brand" (distinguishing elements that are valued by their customers), to entice new customers and increase the loyalty of current customers. The ability to "turn" aircraft quickly from one flight to the next can allow the aircraft to complete more revenue generating flights per day. And, of course adding more seats can clearly create more revenue, but has to be artfully balanced with maintaining passenger comfort within the fixed space of an aircraft interior. So, how does an airline maximize each of these potential revenue enhancing elements? The advent of a thoughtful new range of innovative interior products, can actually help enable all four elements.

1 The Issues

The most straight forward way for cabin interior products to help create more revenue for the airlines is to help them add more seats. In fact, with the creation of new, slimmer seats, which maintain the same comfort at a shorter seat pitch, this has been happening more and more frequently in the last several years. Unfortunately, the flying publics' reaction has noticeably turned negative: "less space", "too crowded", and "no space for my baggage" are now very common complaints. With this in mind, clearly, we also need to address passenger satisfaction, notably in the form of comfort. In order to both add seats and improve or at least maintain

S. Savian (✉)
Zodiac Aerospace, Plaisir Cedex, France
e-mail: Scott.Savian@zodiacaerospace.com

© Springer International Publishing AG 2017
K. Richter, J. Walther (eds.), *Supply Chain Integration Challenges in Commercial Aerospace*, DOI 10.1007/978-3-319-46155-7_3

comfort levels, we need all of the interior components working in harmony to create an optimized *cabin* solution.

Since the products within a cabin interior are typically sourced through multiple suppliers working to specifications which concentrate on optimizing that product alone, this is more difficult than it sounds. Through the Cabin wide introduction of its multiple award-winning Innovative Space Interior System, Zodiac Aerospace pioneered a new generation of space optimizing interior products that work in harmony to create a greater whole. While these products specifically focus on increasing revenue generation (specifically helping to add seats), the real magic in the system is that it places significant focus on *improving* the passenger experience, despite the added seats.

Why continue the efforts to add seats to an environment that is often perceived as overcrowded and uncomfortable already? It is simple math; the value of an added seat is substantial. For instance, weight savings (operating cost reduction) is one of the constant pursuits for aircraft. Not only does saving weight help reduce fuel consumption, but it also has the real benefits of increasing the aircraft range and payload. In fact, it even has a common number to measure the savings generated—approximately $50/lb per year. However, despite the obvious benefits of taking weight out of a flying machine, saving weight pales in comparison to increasing revenue. An additional coach seat on a single aisle aircraft can add up to $75,000 per year in incremental revenue for an operator...so adding a row of six seats can increase annual revenue by $300,000 per aircraft. This is the rough equivalent of reducing weight by 6000 lbs, which from an interior standpoint is a virtually impossible task.

2 The Solution

So, the potential revenue created by adding seats is certainly enticing, but the big questions is will the flying public continue to accept more seats? Through thoughtful integration of passenger cabin elements, both at the component level and at the cabin level, we can produce a more efficient cabin—a cabin capable of creating the most revenue opportunity for the airlines while also *improving the passenger experience*.

Yesterday's aircraft cabin was fundamentally evaluated on cost, weight, and presumably comfort (assuming safety is non-negotiable). Today, the OEMs and airlines use the cabin as a key enabler for both *revenue optimization* and *branding and differentiation*. While revenue generation, in its simplest form, is easily tied to seat count, increasing seats typically reduces comfort and the passenger experience. However, these are exactly the elements that airline branding aims to improve. Traditionally, it was either maximize revenue or improve passenger comfort. Today, we need to do both.

By taking a whole cabin approach to optimization, we can continue to enhance the overall aircraft efficiency, while simultaneously improving the passenger experience. This is fundamentally accomplished in by adding more customer value than we take away. As we take away something a customer values (seat pitch to add more seats), we give them back something else of equal or greater value (such as improved leg room and head room, more features, or improved baggage space). Slimmer seats can provide more leg room, which can be complemented by larger overhead bins which allow baggage that was stored under the seat to be stored overhead. These same bins can also improve overhead space by rotating upwards into the cabin. Additionally we can psychologically improve the cabin spaciousness through improved lighting and materials, while enhancing the experience by reducing the stress and anxiety (i.e. ensuring a space for everybody's bag to eliminate "carry-on bag anxiety"). By taking a holistic view and combining multiple product innovations in a complementary manner, an integrated approach can both increase the revenue potential and improve the passenger experience. The examination below investigates how this integrated approach enabled a major A320 operator to add 12 seats to their cabin, improve their passenger experience, create a true brand, and help to ensure consistent, quick aircraft turns. While the example is specific to a single airframe and customer, the approach can be applied universally.

3 Optimizing the Aircraft

The real estate within a commercial aircraft cabin is among the most valuable in the world. A passenger is typically willing to pay $50 per flight hour for 5 sqft of space. As a quick reference, this would be the equivalent to renting a 1000 sqft (100 sqm) apartment for $7,200,000 per month! How we utilize this space is obviously critical, and the easiest way to maximize the benefit for the airline (by increasing revenue potential) is to add more seats.

With these facts in mind, it is impossible for the airlines and OEMs to ignore the potential of adding seats to their aircraft. However, as noted above, adding more seats traditionally comes at great cost to the passenger. Reduced leg room, less comfortable cushions, less baggage space per passenger, and longer boarding and disembarking are some of the most obvious downsides. The more psychological negatives can be just a critical and are often manifested in areas such as: passenger anxiety for finding baggage space, the feeling of invasion when someone reclines their seat into your lap, and the general feeling that the airlines simply do not care about their passengers. So, the real question becomes, "how do we optimize the revenue capability of the aircraft while simultaneously *improving* the passenger experience?"

4 Passenger Comfort

What is comfortable? Within a single aisle commercial aircraft, comfort can be achieved in many ways: passengers want a comfortable seat, smart entertainment options, a pleasing environment, low anxiety, and a feeling of spaciousness. With the A320 project, the enablers for a better passenger experience were a series of localized product improvements, which, when integrated, created a sum greater than their individual parts. The resulting interior provides more space for the passenger, simple convenience, less stress, and personalized IFE within an elegant passenger space. It all starts with the seat.

4.1 Seats and IFE

The first enabler to help add seats while preserving passenger comfort is the slim seat. These seats, by reducing the thickness of the seatback, provide similar or improved space for the aft passenger at a shorter seat pitch (the spacing between seats, generally, a shorter seat pitch will allow more seats on the plane). The slimmer backs increase space for the aft passenger and provide more room in the critical knee area, as shown in Fig. 1. The slim seats also provide more foot and shin space underneath the seats by virtue of more efficient base structures. Fundamentally, these seats allow seats to be placed closer together while preserving the passenger comfort, but this is far from the whole story.

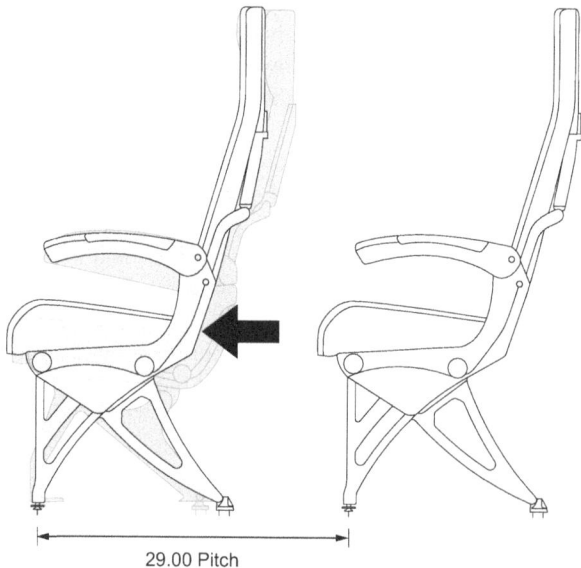

Fig. 1 Slimmer seats create more passenger space, even at tighter seat pitch

The second element in passenger comfort and experience within their seat is the availability of in-flight entertainment. IFE integration into a seat traditionally required a thick, bulky seatback, which either necessitated a longer seat pitch (generally resulting in fewer seats) or compromised passenger space. However, recent advances in IFE technology have made it possible to place wide body type IFE in a slim, single aisle seat (Fig. 2).

Now, the passenger, despite being slightly closer to the other passengers, has equal or improved space and the availability of in flight entertainment. The first building blocks to a better cabin are in place.

4.2 Monuments

The local optimization of the interior monuments, most notably lavatories and galleys, can make additional space available for passenger seats. For the A320, the typical aft lavatories and galley are independent units arranged both fore and aft of the rear doors, as shown in Fig. 3.

The lavatories are sized to support yesterday's common seat pitches (the distance between seat rows...32″, for instance). These monuments, however, are not optimized for use at 29″ pitch, which is a comfortable pitch with today's slimmer seats. By simply changing seat pitch, we oftentimes have not created enough space to add seats in the cabin; the monuments in their standard sizes (lavs and galleys) have become limiting. Here we introduce our next local optimization.

An integrated Lav/Lav/Galley, now known as Spaceflex2, located aft of the rear doors frees up substantial space for seats forward of the rear door. As shown in Fig. 4, combining this integrated monument with the shorter seat pitch creates space for an *additional 12 seats in the cabin* (or an ***increase of 8 %*** for a typical 150 seat aircraft).

Fig. 2 Seatback IFE provides various entertainment options and is now available in slim seats

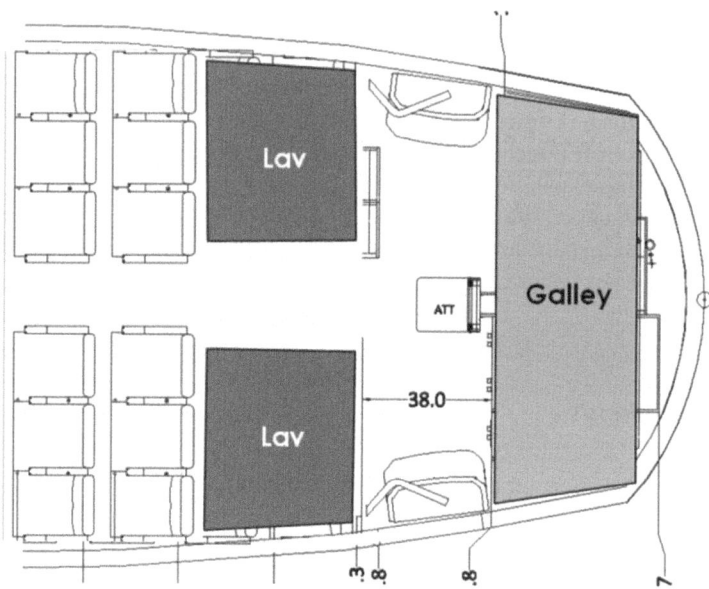

Fig. 3 Conventional aft lavatories and galley arrangement

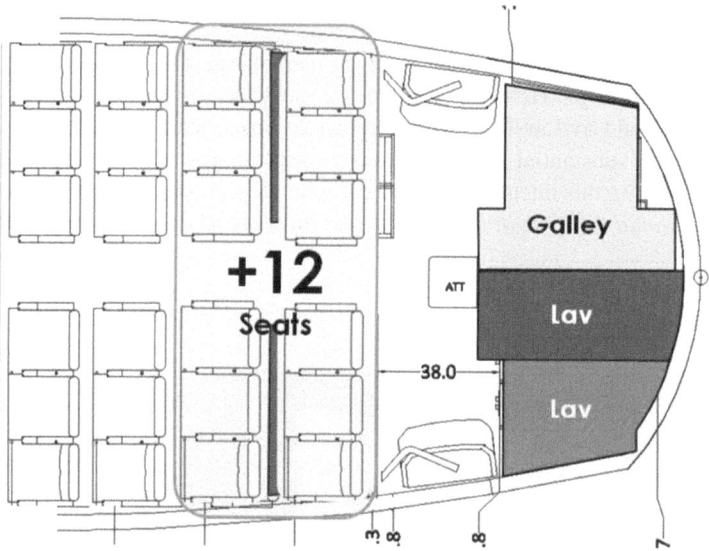

Fig. 4 Optimized lavatories and galley create the space to add 12 passenger seats

This integrated monument includes space optimization within the lavs and galley to provide nearly the same comfort and service capabilities as before, while realizing substantial real estate savings that can now be used for seats (Fig. 5).

Fig. 5 Spaceflex2 with fully functional galley and two full sized lavatories. With optional attendant seat mounted to the inner lavatory door

Working together, the suite of complementary innovative products (slim seat with IFE and the new Spaceflex2 aft complex), have allowed an increase of 12 passengers with a comfortable seat, ample leg room and real entertainment options. A strong start, but we are not quite done.

4.3 Bins, PSE, and Lighting

The additional passengers, albeit with comfortable seating, still means more people in the same space, and those people will also carry on additional hand luggage. Today's aircraft already have major baggage handling issues which negatively impact both the passengers (from feeling of anxiety and frustration to actually incurring additional baggage fees) and the airlines (from unsatisfied customers, excessive work for the flight crews storing bags, and even delayed departures). Adding more passengers only exacerbates this problem. A smarter and larger overhead bin system can solve these problems and also help to improve the overall appearance of the interior.

Where the seat creates the passengers' local environment, the overhead storage bins, passenger service units, the sidewall and ceilings, and cabin lighting create overall cabin environment. At the heart of this environment is the overhead bin.

Today's single aisle bins simply do not hold enough baggage for the today's passengers who prefer the convenience of bringing their bags aboard (approximately 76 % of the flying public, and growing). While we can argue the pros and cons of charging for checked bags, or charging to carry on, the real answer is *to enable all of our passengers to do what they want*. And, what they want, whether they are first on the plane or last, is to bring their bags aboard. We must first solve this fundamental issue.

Where today's conventional fixed shelf (stationary) bin holds up to five bags either wheels first (bag 1, 2, 4, 5) or lengthwise (bag 3), it simply does not allow each passenger to bring aboard their bag. As one can imagine, this creates problems. Passengers who are told at the gate that they must check their bag are unhappy, the crew is put in the undesirable position of having to decide which customers are allowed to do what they want and which are not, and the anxiety around the entire boarding process is stressed. The development of a high volume pivot (moving) bin solution for the A320 provides a better use of available space to directly alleviate this issue. With a design that places a typical carry-on bag into the bin wheels first and *on edge*, the baggage capacity is increased by up to 60 % up to eight bags per bin (bags 1–4 not shown closed bucket), as shown in Fig. 6.

The additional capacity created by this new bin architecture immediately solves the overhead baggage storage issue and clearly provides something that passengers' value. By designing the bins around real life baggage data (for the first time not working to some specification of a theoretical bag which nobody actually uses) and

Fig. 6 Pivoting bins increase carryon baggage capacity

by introducing new patented advancements in bin technology, the bin not only allows more bags, but also larger bags.

The more capable bins ensuring efficient baggage storage is critical, not only in passenger satisfaction, but also in ensuring quick and repeatable turnaround times for the aircraft. In order to schedule as many flights as possible, quick turnaround times (TAT) are essential. The airline must also be able to rely on their TAT to confidently schedule their flights without the potential for compounding delays to affect their last flight of the day, where a missed flight potentially leads to huge costs associated with having an aircraft and crew spend the night in the wrong place.

When it comes to ensuring quick and consistent boarding and disembarking of the passengers (the largest variable in TAT), there is nothing more impactful than making sure each passenger's bag remains on board and is located near their seat. Each time a passenger cannot find a spot for their bag, they begin the inefficient process of looking around the cabin, thereby blocking the smooth flow onto and off of the aircraft. When many passengers have to do this, precious *minutes* are wasted. Further when the passenger and crew cannot find space for a bag (or typically multiple bags), the bags then need to be removed from the cabin *and* placed in the cargo hull, wasting more precious minutes. Several minutes on multiple flights can ultimately mean the cancellation of that critical last flight of the day (missed revenue and increased costs), or becomes limiting on the number of flights an airline will schedule per day (costing revenue opportunity). The larger, more efficient bins virtually eliminate the need to remove bags and dramatically improve the passenger's ability to store the bags at their seat.

Along with the bin, the passenger service unit (consisting of the overhead reading lights, individual air outlets, attendant call button, and Oxygen masks) and the cabin lighting can also have a pronounced impact on the perceived spaciousness of the cabin.

As shown in Fig. 7, the pivoting overhead bins also open up the cabin, creating more space above the seated passenger, more shoulder space in the aisles, and more open sight lines.

By replacing the often dull and dated cabin liners and lighting (particularly yesterday's harsh florescent lighting) the entire cabin environment can become new and feel more open. The advent of LED mood lighting can create a pleasing environment while also allowing the airline to enhance their brand. The A320 project took this a step further to create a signature look by integrating the *cabin lighting into the PSU*. The formerly utilitarian PSU channel now, with the integration of LED mood lighting, becomes an elegant luminaire at each seat row, creating an intimate, passenger-centric environment, as shown in Fig. 8. For the first time, the cabin is lit at the seat row and not the sidewall, allowing for a dramatic visual difference and helping to create a more passenger centric experience.

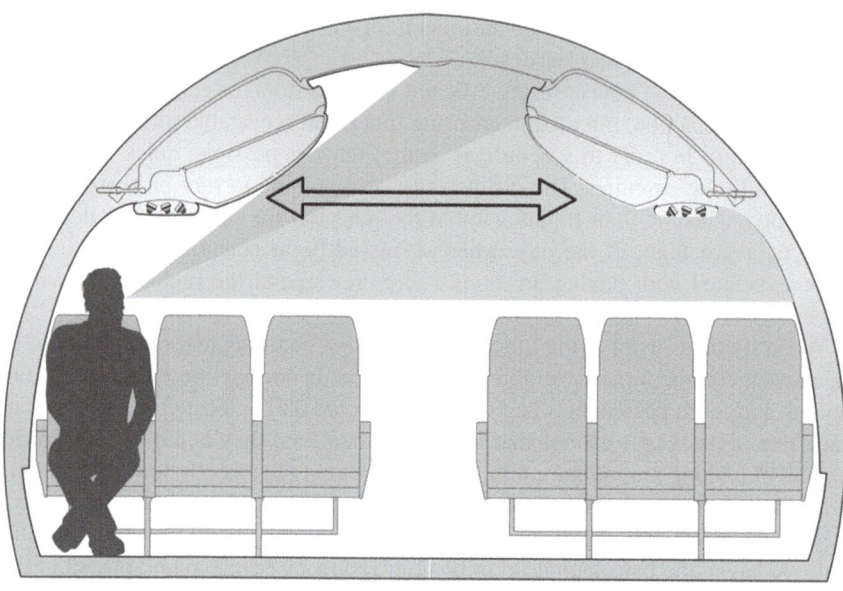

Fig. 7 Pivot bins also increase cabin space above the passengers, improve lines of sight, and widen the aisle

Fig. 8 PSU integrated mood lighting creates an intimate passenger environment and a dramatic cabin

5 Conclusion: The Optimized Cabin

With the A320 project, we can see how an integrated approach can, though a series of thoughtfully designed products, working together, create a whole greater than the sum of the parts. The combined effect of spacious slim seats, inflight entertainment, ample baggage storage, a more open cabin, and elegant details result in a cabin which addresses the four main revenue drivers noted above.

By creating a more enjoyable cabin with improved features, the airline's brand has been enhanced. This paves the way for passenger loyalty and the enticement of new customers, thus helping to increase the percentage of seats sold (the load factor) and improving the yield by potentially selling those seats for more. The larger bins mitigate or even eliminate the baggage problems allowing the airline to execute the most flights per day possible, and the smart combination of slimmer seats and space saving monuments have added 12 seats to the aircraft. After the new interior was in the field for over 6 months, the end result has noted improved customer satisfaction, despite the 12 additional seats.

Interestingly, it took a single supplier to be able to bring this combination to fruition. When integrating an entire cabin with multiple suppliers, each supplier and product spec works independently to ensure that their piece of the interior is optimized. However, local optimization does not always lead to global optimization. A single team based at a single supplier proved it is possible to bring all of the pieces together for a whole greater than the sum of the parts.

While this example is centered around the technologies and products used on the A320 project (as shown in Fig. 9) for a single airline, local product optimization, the integration of complementary products, the basic thought process can be applied to any aircraft for any airline.

Fig. 9 The optimized cabin

Scott Savian is the Executive Vice President of ZEO, the Design and Innovation Studio for Zodiac Aerospace, and has been in the aircraft interiors business for 25 years. He graduated from Worcester Polytechnic Institute with a BS in Mechanical Engineering and received an MBA from the University of Pittsburgh. After moving to California from Boston, he worked at a small design firm, leaving him with an indelible impression of the importance of design and product excellence, before joining the former C&D Aerospace.

Starting as a design engineer, he moved up through the management levels within Zodiac Aerospace, and has been involved with many influential interior programs. In 2000 he moved into Sales and Marketing which lead to a position as EVP of Customer and Product, where he began the plans for ZEO, an integrated design studio that provides a unique, equally weighted, combination of industrial design, advanced concept engineering, and mockup and prototyping. ZEO has since become a major driver of cabin innovation within the industry, with a track record of quickly turning ideas into realized products. In 5 short years, the studio has captured multiple international design awards and helped to introduced innovative products onto nearly 1000 aircraft in service.

Innovation Challenges in the High-Tech, Long-Cycle Jet Engine Business

Alan H. Epstein

Abstract Jet engines are a high-tech, big business with sales of over $50 billion per year. The business requires enormous investments in closely held technology. It is a much different kind of endeavor than any other of today's more common high-tech business that stresses speed to market. Aircraft engines require a very long time commitment since products may take a couple of decades to become profitable, but then continue to generate revenue for many years. Indeed, most of the net revenue is generated by selling parts for and servicing engines produced and sold decades earlier.

The jet engine business is a highly competitive one, dominated by three major companies who have competed for market share using technology and business acumen since the 1950s. The nature of the business has evolved considerably from those heady days at the dawn of the jet age. One significant aspect is the increased importance of the supply chain. In the early days, engine companies manufactured the vast majority of an engine in-house, over 80%. Now the percentages are reversed, with 80% of an engine produced by risk-sharing partners and suppliers. This increased dependence on the supply chain has changed both business and technology aspects of the enterprise. It is a complex chain indeed, with many partners and suppliers working with competing engine manufacturers. For both the manufacturers and suppliers, this broadening of relationships has served to hedge financial risk and provide opportunities to inject new ideas and innovation into products.

1 Introduction

Over the decades, a perceived reduction in product differentiation through technology has changed the market. By the mid-1990s, many thought that aircraft engines had become a commodity, with the main differentiator being the price. This is an

---

A.H. Epstein (✉)
Pratt & Whitney, East Hartford, CT, USA
e-mail: alan.epstein@pw.utc.com

uncomfortable position for a high-tech, long-cycle business. The management of Pratt & Whitney's (P&W's) parent company United Technologies Corporation (UTC), for example, looked for an innovation that could deliver significant customer value to generate product differentiation. The geared, ultra-high bypass ratio engine architecture promised exactly this, so the company committed their financial resources to develop and produce the product. About $1 billion was spent on pre-product launch technology over 25 years. Then another $10 billion in product development, plant capital and sales concessions was needed to bring this innovation to market. The result has been a transformation of the commercial jet engine market within a very few years.

P&W's PurePower® Geared Turbofan™ (GTF) engine stimulated the announcement of seven new or re-engined airplane families, with thousands of these new airplanes quickly ordered. These announced engine orders represent lifetime revenue of several $100 billion of company revenue, with a lot more to follow after the first airplane enters service. For a technical vision to become real, it requires a determined business strategy and carefully calibrated financial planning. As is often the case in the introduction of an innovative product in an established business, overcoming conventional wisdom is a major hurdle.

The *Oxford English Dictionary* defines 'innovation' as "The action of introducing a new product into the market" and "something newly introduced; a novel practice, method, etc." It is distinct from invention. For the innovation to be successful, it must be about more than novelty, it must be about value. Value creation is the core of the Geared Turbofan engine introduction.

2 Engine Innovations

In 1925, many experts believed airplane engines should be liquid-cooled. The air-cooled, radial Wasp engine invented by Pratt & Whitney founder Frederick B. Rentschler, however, was simpler and lighter than any liquid-cooled engine that dominated aviation in those days. This innovative design helped to change aviation.

The Wasp's strength was not that it introduced any particular novel feature that hadn't been tried before. Rather, in C.F. Taylor's words describing the Wasp in his definitive 1971 history of the aircraft piston engine, "While most of these features had appeared previously, their combination here was an eminently rational and successful one, and set a high standard for future development of radial engines" (Taylor 1971). Taylor went on to call the Wasp, "...the first large radial air-cooled engine of what may be called "modern" design" and "the basic features of the Wasp ... are used in all modern large air-cooled radial engines. This type, of course, has dominated transport and much of military aviation until the recent advent of the jet and turbine engine" (Taylor 1971).

In 1911, the British aviation publication *The Aero* opined, "The problem of the aviation engine is purely the combination of power and lightness and reliability." In aviation, achieving high power, light weight *and* reliability is all about getting the details right. The Wasp developed the desired 425 hp on its third test run. Its performance enabled the speed, rate of climb, performance at altitude and reliability that bested countless aircraft performance records. Perhaps most importantly, these engines quickly gained a reputation for exceptional reliability, clearly a virtue for an aircraft engine.

The engine was an immediate success. Charles Lindbergh shattered the American transcontinental speed record in 1930 with his Wasp-powered Lockheed Sirius. Jimmy Doolittle relied on his Wasp to take his Gee Bee aircraft to new speeds. And Emilia Earhart made history with her Wasp-powered Lockheed Electra 10E. Competitors were forced to adopt many of its features to stay competitive.

Then came the jet plane. Up until recently, there were three eras of jet engine architecture. The first was the single-spool technology of the late 1930s and 1940s. These engines took over military fighter planes from piston engines due to their high speed capabilities, but didn't have the efficiency needed for transports. The second era started with the development in 1950–1951 of the dual spool turbojet engine, notably the P&W JT3C that powered early versions of the Boeing 707 and McDonnell Douglas DC-8. These engines offered a 10 % fuel burn improvement over their predecessors, which was enough to start the commercial jet age in 1958 with airplanes that could cross North America and the Atlantic, albeit barely. By 1962, these dual-spool turbojets had evolved into derivative low-bypass ratio engines that powered later versions of the B707 and DC-8, as well as the DC-9 and B-727 aircraft introduced a few years afterward. The low-bypass ratio versions of these engines delivered higher thrust, lower fuel consumption and lower noise. In terms of airline customer value, these derivatives enabled more passengers, greater range and lower cost. The third era of engine architecture started in 1970 with the introduction into service of JT9D-powered, wide-body Boeing 747. This high-bypass ratio turbofan engine offered another 10 % improvement in fuel economy as well as much lower noise. The trans-Pacific range and the unprecedented economics of the wide body aircraft enabled lower fares that fuelled an immense growth in air transportation, especially among leisure travelers. The next 45 years saw the refinement of the high-bypass ratio engine, continuing to reduce fuel burn, noise and maintenance cost. The high-bypass ratio era continued until 2016.

These earlier eras had two things in common: they were all motivated by about a 10 % improvement in engine efficiency, and each was started by an innovative, new engine. In 2016, commercial aviation entered the fourth era of engine architecture, the ultra-high-bypass ratio, geared turbofan engine era, with Pratt & Whitney's PurePower Geared Turbofan engine. This engine family represents an even larger step improvement in efficiency than was the case in previous transitions, delivering a 15–16 % improvement in fuel burn. This gain was so attractive that it motivated the design of new aircraft and the re-engining of several in-production aircraft models. Before the first engine had even entered service, the offer of the GTF engine to airline customers stimulated the order of over 8000 new aircraft. GTF engines have driven such significant change in the aircraft industry because they dramatically bring value to airplanes, to airlines, and to the communities they serve.

The GTF engine's commercial success is principally based on the value of a 15–16 % reduction in fuel burn, saving 15 billion gallons of fuel over its first

10 years which has the potential to double the profits of the world's airlines. The concomitant reduction in CO_2 emissions from this fuel burn improvement is the equivalent of taking three million automobiles off the road. An additional benefit that is extremely important to airports and local communities is a dramatic reduction in noise. On takeoff, the area of the footprint of objectionable noise on the ground is reduced by about three-quarters; on landing the engines are imperceptible since the airframe makes more noise. This chapter is about the innovation challenges during the value creation.

3 Innovation Versus Conventional Wisdom

In the aircraft industry, like many others, there is "conventional wisdom" and there is reality. The two are often the same, but not always. The innovation of the Geared Turbofan engine was introduced in a seemingly mature business, such as gas turbines. Old conventional wisdom was confronted, disputed and, ultimately, routed by new technical realities.

3.1 Geared Engines in Perspective

Conventional wisdom was geared engines would offer no net advantage over direct drive configurations because of their weight, heat generation and unreliability of high power speed-reduction gears. Indeed, geared engine concepts had been around for a long time, without marked success. One of the first geared fan engines was the PW304, developed as part of a secret U.S. military airplane program in the 1950s called Project Suntan. This engine employed an unusual expander cycle since it was fuelled by liquid hydrogen, so it was not a turbofan by today's understanding, but it did have a geared fan. Liquid hydrogen proved to be a poor fuel for high-speed airplanes, so the project was canceled after engines ran on test stands. No examples of this engine are currently known to exist.

In the late 1960s and early 1970s, a number of companies built geared turbofan demonstrator engines, mostly with the objective of lowering engine noise by using gears to enable relatively large, low-speed variable-pitch fans. The gears and the variable-pitch fans of that era were problematical and the engines were unremarkable in terms of performance. These variable-pitch fan engines never saw service. Two geared engines that did enter service employed reduction gears for a different purpose: to convert existing turboshaft engines into turbofan engines at minimum development expense. Other than their use of gears to drive the fan, these engines had little to recommend them compared to their direct drive competitors of the time. One piece of conventional wisdom that resulted from all of these efforts is that fan drive gears were a problem because they generated excessive heat and suffered from reliability problems.

The value that a geared architecture offered was recognized and the experiences of the 1970s were regarded as lessons in the challenges to be overcome. Investments in geared engine technology were renewed in the late 1980s by Pratt & Whitney, in cooperation with NASA. The first public demonstration of progress was the 40,000 lbs thrust P&W Advanced Ducted Propulsor (ADP), run in 1991. The ADP was designed as a technology demonstrator, ground test engine to explore various concepts, including a more advanced variable-pitch fan. This ultra-high-bypass ratio (15:1) engine was the first to demonstrate thrust reversing using a variable-pitch fan. One lesson learned was that the compromises needed to make a variable-pitch fan work negated much of the potential value of the concept. It was clear that both the gear and the fan were the keys to arriving at a design that delivered full customer value. The next decade was spent on a focused investment in flight weight gear technology. In 2001, a flight weight gear system and fixed-pitch fan in the 11,000 lbs thrust P&W Advanced Flight Technology Integrator (ATFI) engine was demonstrated.

All of this investment came to a critical point in the first decade of the twenty-first century. The favorable results from the AFTI engine encouraged the funding of a large gear test rig that was needed to demonstrate gear endurance and flight-representative loads. Finally in 2007 and 2008, a 24,000 lbs thrust demonstrator engine encompassing all of the learning to date was built and tested and then flown on the company's Boeing 747 test aircraft. This engine was quickly shipped to Airbus, who wrung out the concept on its A340 airliner. This engine demonstrated that the investments in gear technology had paid off. The gear generated less than 25 % of the heat that would have been expected from the early technology demonstrators or from geared or turboprop engines then in service. The high-bypass ratio fan demonstrated a step reduction in noise levels. By this point, about a billion dollars had been invested in technology maturation. The bill for product development was still to come.

This extensive testing, particularly the flight testing, was sufficient to convince airframe manufacturers and airline customers that the geared turbofan architecture such as the PurePower Geared Turbofan engine (Fig. 1) was indeed a very promising approach. Within a short time, first Mitsubishi Regional Jet (MRJ), then Bombardier, then Airbus, then Irkut, and finally Embraer all ordered GTF engines for their new or re-engined airplanes. The 15–16 % reduction in fuel burn was just irresistible.

The GTF engine's success is not just about the gear, however. The perfection of a gear enabled new engine architectures introduced new challenges and opportunities for engine designers as well as for partners and suppliers. About 50 new technologies were managed and matured in order to transform an innovative advanced concept into a low technical-risk, attractive product. Let's examine a few of the most important technologies to illustrate how innovation and investment overcame conventional wisdom, including roles played by the supply chain.

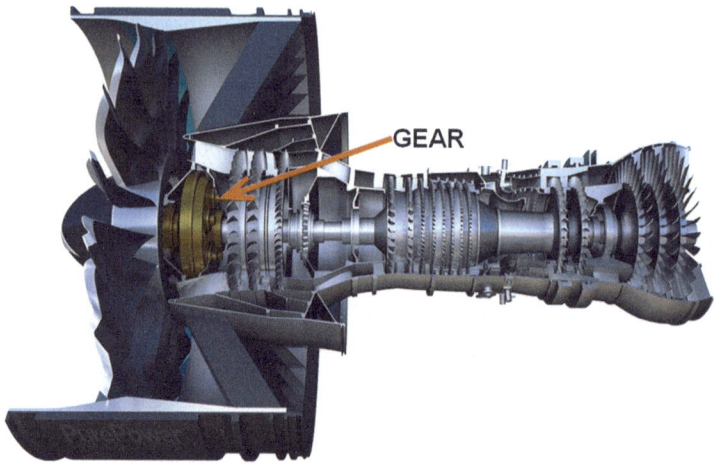

Fig. 1 The P&W Geared Turbofan engine, gear system highlighted, 2010. ©2016 Copyright United Technologies Corporation—Pratt & Whitney Division

3.2 Fan Drive Gears

Gears were considered to be too heavy, had too much waste heat to reject and their lives would not be long enough to be economically viable. The requirements were clear, yet extraordinary by historical standards—efficiency better than 99½%, a lightweight but very rugged design, and very high reliability. The GTF engine fan drive gear system is a feat of advanced mechanical engineering. It transmits up to 25 MW of power in a package less than 40 cm in diameter, while weighing only slightly more than 100 kg. It needs maintenance only every 20 years. It is deceptively simple, requiring only 13 major parts. It is also extraordinarily rugged—capable of surviving such extreme, if highly unlikely, events as bird strike or fan blade loss. This success is based on getting all of the design details right. Getting there may have required a prolonged, even prodigious investments but all of these requirements were achieved. And, this achievement has changed the face of commercial engine design.

3.3 Variable-Pitch or Variable Area Nozzle

Low-pressure ratio fans were required for an engine to be efficient. Its fan should produce relatively little pressure rise so that the velocity of the engine's exhaust jet is low, which also helps to reduce noise. It was widely believed that at fan pressure ratios much below those in service at the time, fans would need either variable-pitch or a variable area fan nozzle. Indeed, those demonstrator engines in the 1960s and 1970s had variable-pitch, but neither variable-pitch nor a variable area fan nozzle had ever been introduced into service. In fact, as late as 2010, the idea that variable-pitch or variable area nozzles were needed was widely appreciated and recognized in the industry (Cumpsty 2010).

In other words, a variable area fan nozzle was seen as an enabling technology for lower fan pressure ratio. P&W worked extensively with its nacelle supplier to develop what was to be the world's first commercial aircraft variable area fan nozzle. The challenge was to provide the functionality required at low enough cost, weight and perceived technical and operational risk that it would be readily accepted by customers, cautious of too much innovation. If you look at the demonstration engine flown in 2008, you'll see that it had a variable area fan nozzle, the first ever flown on a commercial turbofan. And yet, none of the geared turbofans now going into service have variable area nozzles. Why? Because extensive testing of this new and innovative low-speed fan showed that "conventional wisdom" was wrong. The unprecedentedly low fan speeds combined with the fruits of modern computational tools yielded a fan design robust to the instabilities that plagued legacy designs, instabilities that the variable area nozzle or variable pitch were intended to avoid. The GTF engine fans operated perfectly well without the extra protection provided by a variable area nozzle. By eliminating the nozzle's variable area capability, nacelle weight and cost could be reduced, so off it came for production.

3.4 Large Diameter Nacelles

Even if the geared engine worked, the nacelles required for these geared turbofans would be too big—too large in diameter, too heavy—and would suffer excessive drag in flight. One exemplary challenge was, "How do you take an airplane such as the Airbus A320, highly optimized for a 63-in. diameter fan, and fit an engine with an 81-in. diameter geared fan under it?" The value generated by 16 % fuel burn improvement was enticing. As can be seen in Fig. 2, one can indeed fit a significantly larger fan under a wing when sufficiently motivated. One enabler was to design a thinner nacelle so that more of the total outside diameter could be allocated to the fan. This required working closely with both the nacelle supplier and the suppliers of the engine accessories that fill-up much the nacelle volume surrounding the engine.

The larger diameter nacelle may fit but what about its weight and drag? With the advanced technology incorporated into the GTF, the diameter of the nacelle could grow appreciably, but its length has been dramatically reduced, minimizing weight and drag penalties. Figure 3 shows the PW1100G-JM engine for the Airbus A320neo engine on the Boeing 747 test airplane's inboard pylon. The GTF engine is about a third the length of the legacy B747 engine on the outboard pylon, even though it looks to be similar in diameter. Why is the new engine so short? One reason is noise, or the lack of it. The inlet lengths on these new GTF engines can be about half those of legacy engines because the slow turning fan is so quiet. Much of

Fig. 2 A composite of an Airbus A320, with a GTF engine on right, V2500 engine on left (Courtesy of JetBlue)

the inlet is there to provide noise attenuation. The geared turbofan engine's slowly turning fan produces very little noise to begin with, so less attenuation is needed. Thus both the inlet and exhaust ducts can be significantly shorter, resulting in less weight and drag.

The engine within the nacelle is shorter too, another benefit of the geared approach. The gear enables slowing the fan, but it also facilitates speeding up the rest of the low-pressure spool, enabling a significant number of stages to be removed. As a result, the engine is shorter, the nacelle is shorter and additional weight is saved. Indeed, on the A320neo the geared engine is several hundred kilograms lighter at the airplane level compared to a direct-drive alternative.

3.5 Fan Blade Construction

Fan blade material is an important discriminator since it has a multiplier effect on engine weight and dynamics. As weight savings, solid titanium fan blades were introduced in the JT3D engine for the Boeing 707 in the early 1960s. Advanced, hollow titanium fan blades took the lead in the 1990s on the Boeing 777. By the turn of this century, "conventional wisdom" held that plastic composite fan blades were

Fig. 3 PW1100G engine (inboard) on the Flying Test Bed with JT9D engines (Pratt & Whitney)

the future—a superior, lighter solution than hollow titanium. Many believed that a 3-D woven, fiber-reinforced composite fan blade was needed for new engines.

Such blade was developed, but a small group of engineers recognized that due to unprecedented low rotational tip speed of a geared turbofan and improvements in metal alloys, other lightweight blade material approaches were viable. The fan blades on the A320neo and the Bombardier CSeries engines are a composite, but one based on metals rather than polymers. These blades are a metallic hybrid with a titanium leading edge and an aluminum body. Realization of the required material properties and quality required working closely with the material and forging suppliers. This design solution wound up being lighter and less expensive than 3D woven composite blades. The real deciding factor was, for this size engine, the hybrid metallic blade could be thinner to improve airflow and efficiency by about 1 %.

3.6 Combustors

Jet engine combustors are extremely efficient, well over 99 %. To a large degree, it is the need to reduce regulated emissions, such as nitrogen oxides (NOx) and smoke, that drives technology investment in combustors. NOx, in particular, has proven challenging. There are two known approaches to NOx reduction, lean-burn and rich-burn-quick-quench. Starting in the early 1990s, there was a consensus the lean-burn approach would be the only path to achieving even lower emissions in future, although it added weight, complexity and maintenance burden. Yet, here we are 25 years later, and the GTF engines feature a simple, elegant rich-burn-quick-quench combustor. A third-generation design, the TALON™ X combustion system offers extremely low emissions and smoke levels—as low as anything in its class (it actually surpasses the most stringent standard for NOx emissions by 35 %). It is highly durable and much less complex than lean burn approaches. Also, it does not suffer from in-flight blow-out and restarting problems characteristic of lean combustors.

3.7 Turbomachinery

The most direct design opportunity afforded by the availability of a breakthrough gear was to better optimize the engine's rotating machinery. The addition of a gear turns a two-shaft engine into a three-shaft engine. This provides the opportunity to drop the fan speed relative to that of the low pressure spool, improving efficiency and weight. But, more shafts mean more bearings and, thus, added complexity if done in a conventional manner. In this case, working with the supply chain produced an innovative bearing that enabled a simplified design.

There was a widespread belief that higher-pressure ratio core compressors were superior to those with lower pressure ratios. This mindset is the legacy of direct-drive two-spool engines that dominated engine architecture for the preceding five decades. Of course, for engine performance, it is only the overall pressure ratio of the engine's compressors that is important, not how it is distributed among compressors. The superiority of higher pressure ratio on the high-pressure compressor is certainly not the case for three spool engines, where relatively little pressure-rise is done on the high spool, only 7–9:1 compared to the 16–23:1 on direct-drive two-spool engines. Historically, both two- and three-spool engines have been equally successful commercially. The optimum high-compressor pressure ratio is a function of the engine architecture. So it should surprise no one that the new two-spool, geared architecture of the GTF engine has its own optimum, one that lies between those of the direct drive two-spool and three-spool engines.

The geared architecture enables a much higher speed, low-pressure spool so that unlike a direct drive engine, there is no weight penalty in moving compression from the high-pressure to the low-pressure spool. In fact, there is a significant benefit. First, low-pressure turbines general operate at higher efficiency than do high-pressure turbines because the low-pressure turbines need not be cooled, thus avoiding cooling losses. Second, by operating at higher rotational speed, the geared low-pressure turbine gains significant aerodynamic efficiency compared to its direct drive equivalent. Together, these mean that geared engine efficiency can be improved by moving compression from the high- to the low-pressure spools.

The turbomachinery in a geared engine also enjoys significant weight, length and complexity advantages compared to the direct drive approach. Since the work done by turbomachinery scales with the square of the rotational speed, a high speed low spool, such as in GTF engines, needs fewer stages. The low-pressure turbine on the geared turbofan engine has three stages compared to the seven stages of a direct-drive design. This significantly reduces engine weight and length, and also cost, since there are fewer, smaller parts. The high-speed low-pressure compressor also saves stages. The savings in low-pressure turbine stages alone more than compensate for the weight and cost of the gear driving the fan. Adding it all up, the geared engine has 40 % fewer airfoils than its direct-drive equivalent. This has several implications for the supply chain. First, almost all of the compressor stages are now integrally-bladed-rotors (IBRs) in which the blades and disks are machine from a single forging. Thus, for example, 50 parts in an old design become one in the new design. This implies a shift in the supply chain as traditional airfoil and disk suppliers must step up to become IBR manufacturer or be supplanted. Also, the significant reduction in part counts will be reflected in factory loads.

3.8 The Whole Versus the Sum of the Parts

So far, we have discussed some of the components of the geared turbofan engine separately, but an engine can be more than just the sum of its parts. This is certainly the case with the GTF engine where an overall vision of the architecture facilitated a synergistic integration, realizing the best of the components.

Incorporation of a fan drive gear allows the use of a high-speed low-pressure spool, dramatically reducing its weight while gaining efficiency. This weight savings was then invested in a large diameter, very slow speed fan to gain efficiency and reduce noise. The low noise and stability of the large, slow fan permitted a shortened inlet and nacelle to reduce nacelle weight and drag. All of these technologies are necessary; none by itself is sufficient. The whole concept would be at risk if one of these links had failed to deliver. Thus, the GTF engine breakthrough required all of these technology innovations to be ready before the launch of the development program to provide manageable business risks. The billion dollars invested in technology readiness proved to be money well spent.

4 Business Value from Technology Innovation

Two of the new aircraft are now certified and in service, the A320 and the CSeries, with three more to follow. The GTF engines provide thrust in the 17,000–35,000 lbs class (with future growth to 40,000 or more pounds), for new generation and re-engined single-aisle and regional aircraft. In early 2016, the Airbus A320neo jetliner was the first aircraft to enter service with this new technology, powered by PW1100G-JM engine. The Bombardier CSeries aircraft followed a few months later. The Mitsubishi Regional Jet, the Irkut MC-21, and the Embraer E-Jet E2 will enter service in the 2017–2019 period (Figs. 4, 5 and 6).

Since fuel is a major cost of commercial airlines, the GTF engines' reduction of 16 % in fuel costs saves airlines significantly more than $1 million per aircraft each year. Its low fuel consumption translates to savings of 11 billion gallons of fuel by

Fig. 4 CS100 (Bombardier)

Fig. 5 A320neo (Airbus)

2025, worth some $39 billion (based on a projected average fuel price of $3.50 per gallon).

The GTF engine is a "green" powerplant system, featuring significantly lower emissions than more traditional engines, with a reduction of regulated emissions of more than 50 % and fleet carbon emissions reduced by some 106 million tons, or more than 3500 tons per aircraft per year, a level equivalent to planting nearly one million trees or removing three million cars off the road, every year. Manufacturing

Fig. 6 MRJ90 (©2016 Mitsubishi Aircraft Corporation)

is keeping the environment in mind, with the goal of minimizing the environmental footprint of the PurePower family of engines throughout the product life cycle. In designing the engine, particular attention has been paid to eliminating environmentally undesirable materials in manufacturing and repair processes. The geared architecture reduces the noise footprint by more than 75 %, well below ICAO Stage V noise standards. The low noise signature also offers the potential to optimize airplane takeoff trajectories, saving up to 2 min of flight time when the engine is operating at maximum power.

In addition to fuel savings and environmental friendliness, the GTF engine's technologies and designs were chosen with maintenance in mind. Engine maintenance is a major airline expense, with overhaul being most of the engine related cost. The GTF engine's full-authority digital engine control (FADEC) and health monitoring provide exceptional prognostics capability. These and other new technologies will extend time on wing between overhauls. Overall, the GTF engine requires 40 % fewer maintenance procedures and inspections.

The benefits of this engine architecture are now so obvious and overwhelming that all new commercial engines will utilize the geared architecture for decades to come. The concept has enormous headroom to grow, both in size and performance.

References

Aero Engines (1911) The Aero, p 44
Cumpsty NA (2010) Preparing for the future: reducing gas turbine environmental impact. J Turb 132(4):041017-1–041017-17
Fayette Taylor C (1971) A review of the evolution of aircraft piston engines. Smithson Ann Flight 1 (4):1–137

Alan H. Epstein, Dr., is Vice President of Technology and Environment at Pratt & Whitney, a United Technologies Corp. (UTC) company and a world leader in design, manufacture and service of aircraft engines and auxiliary power units. Epstein is responsible for setting the direction for and coordinating technology across the company as it applies to product performance and environmental impact. He leads Pratt & Whitney's efforts to identify and evaluate new methods to improve engine performance and fuel efficiency for all new Pratt & Whitney products. He also provides strategic leadership in the investment, development and incorporation of technologies that reduce the environmental impact of Pratt & Whitney's worldwide products and services.

Prior to joining Pratt & Whitney, Epstein was the R.C. Maclaurin Professor of Aeronautics and Astronautics at the Massachusetts Institute of Technology (MIT) and currently holds an appointment there as professor emeritus. He was also the director of the MIT Gas Turbine Laboratory. His research and teaching while at MIT was concerned with gas turbines, power and energy, aerospace propulsion and micromechanical and electrical systems (MEMS).

Epstein has more than 140 technical publications and has given about 200 plenary, keynote and invited lectures around the world. He has won international awards for his work in heat transfer, turbo-machinery, instrumentation and controls, gas turbine technology and MEMS. This includes four American Society of Mechanical Engineers (ASME) International Gas Turbine Institute (IGTI) best paper awards, the Aircraft Engine Technology Award and the Gas Turbine Award. He was the ASME IGTI Gas Turbine Scholar in 2003.

Epstein is a member of the U.S. National Academy of Engineering and a fellow of the American Institute of Aeronautics and Astronautics and the American Society of Mechanical Engineers. He received his B.S., M.S. and Ph.D. degrees from the Massachusetts Institute of Technology in aeronautics and astronautics.

Open Innovation in the Aviation Sector

Johannes Walther and Daniel Wäldchen

Abstract This article examines to which extent the two global aircraft manufacturers Airbus and Boeing have employed open innovation, and how it has impacted on their performance outcomes. This paper provides the theoretical concept of open innovation and an outline of major relevant trends for the aviation sector to underpin the appraisal of the industry. The following chapter introduces the methodology selected for researching the companies' strategies. The findings related to open innovation in the two companies are presented, enabling conclusions to be made regarding the employment of this strategic tool and its effectiveness.

1 Introduction

The commercial aviation sector is defined as a sector of the aerospace and defense industry, which is experiencing transformational change. This is due to a change in emphasis for growth opportunities, from the military to the commercial sector (Deloitte 2015). Long and complex supply chains with Original Equipment Manufacturers (OEMs) characterize the aviation sector. In order to meet contract deadlines within the forecast margins, global companies such as Airbus, Boeing and Bombardier depend on these organizations to deliver components and services that meet the required quality standards in a timely, cost effective manner (CG 2011). The complexity of the supply chain is illustrated in Fig. 1.

The aircraft manufacturers (OEMs) are experiencing declining profit margins while having to face the political and environmental challenges of optimizing fuel efficiency and minimizing emissions (Deloitte 2015). In order to reduce costs, some manufacturers have moved production to lower cost locations and they are outsourcing design-and-build-packages of construction components to their suppliers, instead of assigning build-to-print orders to them (Lewandowska 2012). Hence, OEMs are forecasted to become increasingly reliant on various types of

J. Walther • D. Wäldchen (✉)
IPM AG, Hanover, Germany
e-mail: d.w@ipm.ag

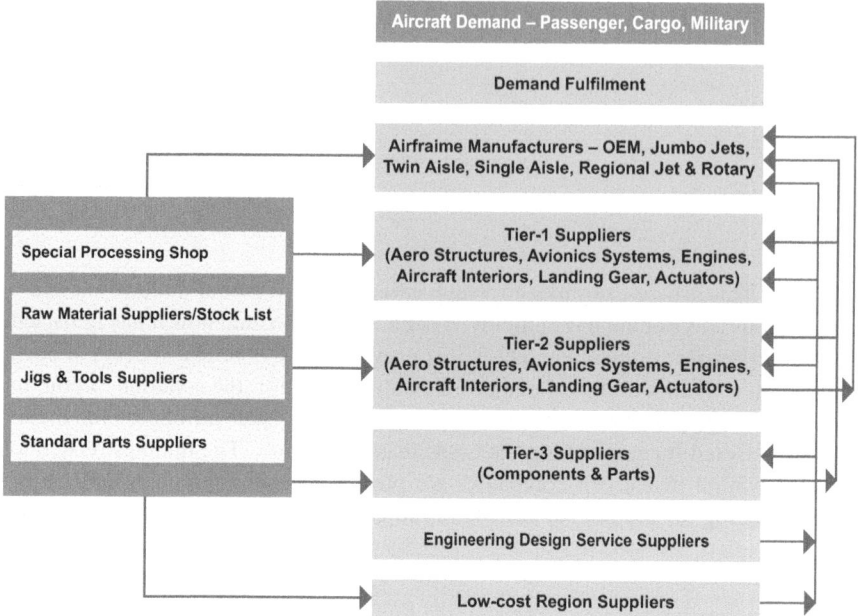

Fig. 1 The aerospace and defense supply chain. Source: CG (2011)

suppliers within the supply chain. They need to find innovative ways of meeting the technological challenges in order to retain profit levels and meet the growing demand for new commercial aircraft (KPMG 2013; PRN 2015).

Both Airbus and Boeing have almost doubled their monthly output of aircraft over the last 10 years (OW 2015). Several companies, among them GE, have used open innovation concepts to find new and effective ideas to meet some of the challenges in production (Chesbrough 2003a, b). The advantages of open innovation are explained in the following chapter.

2 Theoretical Background

Innovation has always been a keyword for describing commercial success. Innovation is seen as the heart of any business activity, as in the following definitions: "Innovation is widely considered as the life blood of corporate survival and growth". "Innovation represents the core renewal process in any organization. Unless it changes what it offers the world and the way in which it creates and delivers those offerings it risks its survival and growth prospects" (both quoted in Baregheh et al. 2009).

Innovation can be seen as a process in which basic research leads to creative ideas which are commercialized by making the appropriate investments. However, this can be a long, complex route, and there is no certainty of commercial success (DIUS 2008). Therefore, many companies are basically making creative changes to established products and services by employing new technologies.

The need for innovation may come from different directions, though. Changing lifestyles and changing public demand may sideline one category of products and increase demand for another kind of product. In the automotive and aviation sector alike, the demand for less fuel consumption is adamant. Innovation is therefore driven by the constant demand for new ideas which has led some organizations to look beyond their internal resources to gather ideas from universities, suppliers, competitors and other organizations (DIUS 2008). In contrast to traditional in-house research and development, this innovation process is open rather than closed. Even though the idea is not a particularly new one, Chesbrough (2003a, b) popularized it when he propagated what he considered a new mindset within the sector.

Closed innovation is described as the process in which a firm makes all the decisions and choices regarding the development of its products and services itself. Whereas open innovation is characterized by seeking and incorporating ideas and/or components generated by external contacts (Almirall and Casadesus-Masanell 2010). In large organizations, closed innovation is typified by an isolated centralized research and development division generating new products, which the company finances, manufactures, distributes and sells. These organizations have the responsibility for the whole vertically integrated supply chain (Chesbrough 2003a, b). Hence unique knowledge is retained within each company, and employment of new talent is the major intervention that could lead to a generation of different ideas. According to Chesbrough (2003a, b), however, changing environmental factors led to an erosion of closed innovation as the standard industry model:

- Labor mobility resulted in competitors gaining new ideas.
- Availability of venture capital facilitated talented employees to move to a competitor.
- Research ideas were left undeveloped, motivating researchers to create competing start-up companies to commercialize their ideas.
- Emerging supplier capability allowed to transfer R&D ideas from one company to another, facilitating supply to competitors, too.

The pace of open innovation adoption has been driven by shorter innovation cycles, resource scarcity and the rising cost of in-house R&D (Gassmann and Enkel 2004). Despite huge knowledge banks inside companies, the level of innovation may not be equivalent. The advantage of adopting an open innovation strategy is that firms access ideas that would never have been generated internally, according to Almirall and Casadesus-Masanell (2010). Companies can make non-incremental changes to their product portfolio, applying diverse knowledge of their sources to leverage their competitive advantage. There is substantial evidence, however, that companies which continue to employ closed innovation strategies often fail in delivering truly innovative products, owing to the incapacity to detect changes in the environment (Chesbrough 2003a, b).

Open innovation is defined as an interactive process between company and market (Hilgers and Piller 2009). The aim of cooperative innovation management is to enhance the effectiveness and efficiency of the innovation process. This requires a high innovation competence, which can be considerably increased by integrating external solutions, in particular in the early stages of the development process (Reichwald and Piller 2009).

Providing access to external information will extend the corporate know-how and provide sales market relevant information on customer demand in the early planning stages of the development process (Hilgers and Piller 2009). In this way, forecast customer requirements can be identified at an early stage, which may lead to new products and services which can be placed in the market at an early stage (diffusion of innovation). The Internet-based methods of co-operative innovation management include innovation communities, innovation toolkits, innovation market, and innovation competitions (Habicht et al. 2011).

A main challenge is to employ the right method at the right phase of the innovation process. This requires a defined innovation process, where internal and external (new) knowledge are linked to the already existing know-how. Figure 2 below shows that external knowledge may impact on an entire company and on all fields of knowledge generation.

GE Aviation, also an important supplier of aircraft components, was looking for an alternative manufacturing process of aviation castings and tried to speed up aviation parts inspection and to increase its accuracy. The company invited contributions through a global challenge. 89 participants tackled the speeding up inspection issue and contributed a diverse range of solutions. 51 entries came from

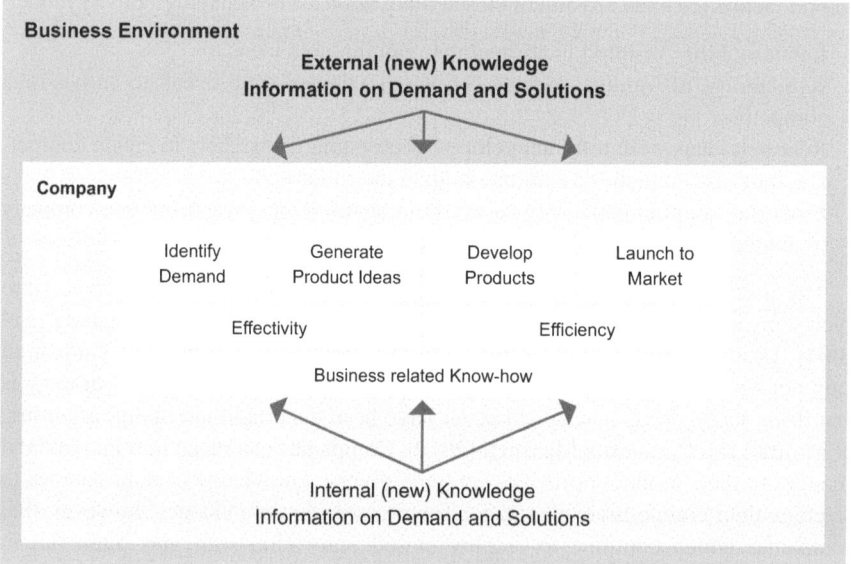

Fig. 2 Open innovation process

48 participants in 18 countries to address the second issue to increase accuracy. The entire challenge showed GE many new approach options and introduced them to a high number of hitherto unfamiliar companies (GE Aviation 2015).

"The challenge was a fantastic way to broaden our exposure to potential new players and new technologies. Only two of the organizations that participated were previously known to us, and it's inspiring—and eye-opening—to see such a broad variety of submissions from around the globe", commented GE's chief manufacturing engineer (GE Aviation 2015).

A study by Gassmann and Enkel (2004) investigating 124 companies found that there are three major models of adopting open innovation: an outside-in process, an inside-out format, and a coupled process of outside-in and inside-out procedures. The relationship between these is clarified in Fig. 3 below.

The outside-in process comprises collaboration of customer and supplier knowledge, in order to supplement in-house expertise, to increase innovation and competitiveness. This can be accomplished in several ways, such as buying intellectual property, being a participant in innovation clusters to increase technical knowledge or by creating joint ventures.

The inside-out process involves: exploiting the firm's ideas externally in different market sectors from its own, for instance, by selling the firm's intellectual property or technologies. In this way, research ideas are more swiftly developed and reach commercialization stage more quickly as if internal means were used only. Outsourcing is another form of this model: the firm shares its knowledge with other firms within its own sector, for reasons such as lack of in-house expertise, poorer expertise than is typical for the company to which the process is outsourced to, speed, capacity management or cost benefits. Licensing patents is both a source of income and a method of sharing innovations that result from the licensee's use of intellectual property.

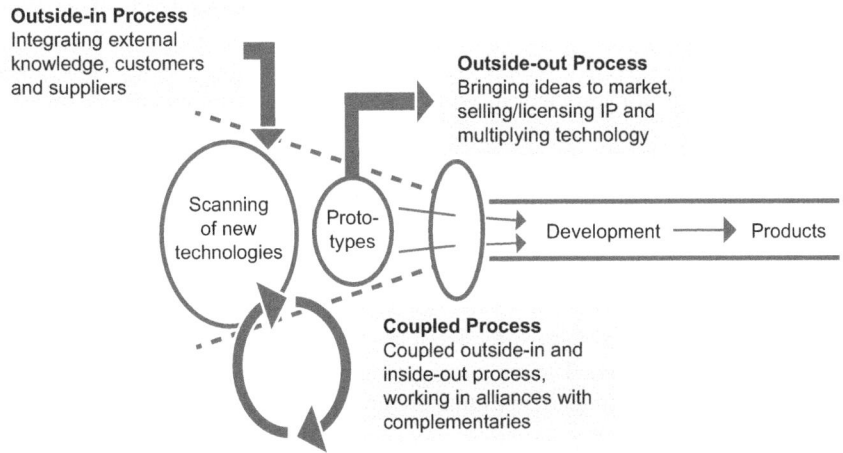

Fig. 3 Three models of open innovation. Source: Gassmann and Enkel (2004)

The coupled model is characterized by a formation of alliances with organizations in the same sector. The crucial success factors are: to gain new knowledge and to transform ideas to commercial market offerings by mutual collaboration. This might take the form of a cooperative arrangement between competitors (Gassmann and Enkel 2004). The initial arrangement between Chrysler and Fiat was such a collaboration: Chrysler employed the fuel efficient engines developed by Fiat to survive in the US automobile market, whilst giving Fiat a stake of Chrysler, and raising its global profile as an innovative company that is able to support innovation to various industries (Di Minin et al. 2010). These models are important in appraising the open innovation direction preferred by the aircraft manufacturers that are considered in this article.

According to Gassmann et al. (2010), a variety of research approaches have been developed:

- The spatial perspective: organizations position their sites at recognized centers of excellence for their section, in order to gain access to the best relevant talent.
- The structural perspective: the entire innovation task is divided and the value chain is tightened and optimized in order to increase specialization and reduce costs.
- The user perspective: users are involved in the innovation process at an early stage to meet their perceived preferences.
- The supplier perspective: suppliers are integrated in the innovation process to leverage innovation outcomes.
- The leverage perspective: companies sell their intellectual property to diversify risk and create new revenue streams.
- The process perspective: firms use three distinct processes, namely outside-in, inside-out and coupled processes.
- The tool perspective: organizations provide tools for users to input ideas or solve issues.
- The institutional perspective: companies enable others to use innovations without license.
- The cultural perspective: companies create a mindset that values open innovation practice, for instance, incentive schemes and management information systems.

The value streams that the commercial aviation sector has adopted can be identified according to these perspectives, and its open innovation effectiveness can be appraised accordingly (Gassmann et al. 2010).

In the automotive sector, OEMs have traditionally innovated by means of their in-house R&D. The study of five large OEMs from Ili et al. (2010) demonstrates that open innovation was more effective than closed innovation in terms of generating R&D productivity. The pressure to innovate more rapidly and to retain margins as costs rise are challenges that are unlikely to be met by a closed model. A similar situation has been experienced by the aviation sector as the overview of the aviation sector trends demonstrates.

Despite extensive studies which promote open innovation, there are potential disadvantages, particularly when the central organization has a network of partner relationships and wants to have all rights to the intellectual property of the suppliers or partners who contributed innovative ideas to the final product or service. Such open source innovation would need contractual terms which will be difficult to negotiate (Lindegaard 2010). The speed and cost of adopting an open innovation approach is also an issue when employed for the development of new products, according to Knudsen and Mortensen (2011).

The major trends in the aviation sector result from political, economic, social, technological, legal and environmental/ethical factors, referred to as PESTLE (Johnson et al. 2013).

The political factors comprise:

- Increasing global political tensions generate demand for military aircraft.
- Regulations force OEMs to develop lower emission models.
- Governments hold back or enforce the sector's consolidation (PRN 2015; OW 2015).

The economic factors include:

- Higher commercial demand for new aircraft to replace obsolete models.
- Cost cutting has to be achieved by putting pressure on suppliers to reduce prices.
- Risk sharing will be imposed, especially on tier-1 suppliers (Deloitte 2015; PRN 2015).
- OEMs move more production to emerging economies, for lower labor rates, market access or both (Lewandowska 2012; OW 2015).
- Rising raw material cost due to the requirement to use lighter materials (composites instead of traditional aluminum) leads to higher development and tool costs (CG 2011).

Key social trends are

- Rising labor costs.
- Technical and administrative skills are needed to develop new design modules.
- R&D capability is required to apply the new materials appropriately.
- Supply chain monitoring is needed to ensure that deadlines are kept and costs are restricted (OW 2015).
- Supplier consolidation is supported by Government Development Agencies in some countries.
- OEMs require tier-1 suppliers to take responsibility for performance and innovation in the lower tiers of the supply chain (PWC 2012).

The major technology trends are:

- Suppliers are asked to provide innovative products and ideas to the OEMs.
- Innovations mainly serve to reduce production cost, reduce operating cost or improve passenger comfort.
- Constant search for improved aircraft safety to lessen corrosion and fatigue.

- Composite materials development promises high cost savings in the long term (OW 2015).

A vital part of OEMs triple strategy of cost cutting, risk reduction and innovation is the altering of contracts with suppliers—a method that increases supplier consolidation (OW 2015).

3 Methodology

The research method employed in this paper is pragmatic. This approach allows an external view on the subject, provided by multiple views expressed in the reviewed research literature. Therefore, it is the best way to research two competing major aircraft manufacturers, since it means that observable phenomena and subjective meanings are equally used to gain knowledge dependent upon the intent of the research (Saunders et al. 2016).

Generally, the pragmatic method focuses on practical, applied research. In this case, the focus is on the research and development processes of two leading companies. The process by which the companies gain new ideas, concepts and methods provides meaningful insights into the nature of innovation (Saunders et al. 2016).

The method requires for the researcher to assess data very carefully. Different, possibly conflicting perspectives must be integrated to successfully interpret the data. Thus, the researcher must be aware that values play a large role in interpreting the results. It is possible that the researcher adopts both objective and subjective points of view. Hence, mixed or multiple method designs can be employed in pragmatic research, using quantitative and qualitative data alike (Saunders et al. 2016).

This paper uses a case study of two companies for various reasons. The case study strategy considers empirical phenomena as evidence, it seeks to answer questions such as how, why and what, and can use multiple sources of data and evidence. In addition, gaining deep insight into a small number of organizations is the key purpose of this strategy. It mainly relies on qualitative data.

Secondary data are employed to appraise open innovation in the two chosen aircraft manufacturer. Of course, there is the limitation of secondary data that the content of the research or the process of data collection can be questionable. In order to minimize this effect, data will be collected from robust sources such as journal articles, official reports, sector magazines, and quality newspapers and websites (Saunders et al. 2016).

The secondary data is subject to qualitative content analysis, which allows for conflicting opinions regarding the question, how open innovation is applied in aviation companies. The rigor of the research is measured by its reliability and validity. In qualitative research, reliability is related to the transparency of the whole process and the validity of the systematic collection and interpretation of the data (Meyer 2006).

4 Findings and Analysis

The findings regarding the question how open innovation is employed at two aircraft manufacturers are presented in this section, commencing with a brief overview of each company, and continuing with examples of open innovation approaches in current use. This will enable qualitative conclusions to be made as to the open innovation models employed by each OEM.

4.1 Boeing

Boeing was founded in 1916, soon after several fatal accidents with planes built by the Wright brothers had cost 11 lives (Boeing 2016a; Crouch 2003). The company quickly supplied aircraft to the US Army, since the US had entered into World War I in 1917. Boeing has supplied the US troops ever since, producing the B17 Flying Fortress and the B52 Stratofortress bombers, among other aircraft. In 1958, the company started delivery of the first US jet airliner, the Boeing 707. Converted into a cargo plane under the name of C40, the 707 laid the foundation for Boeing's ongoing role as a major aircraft manufacturer. The 707 was also to become the first presidential airplane with the moniker Air Force One.

4.2 Airbus

Airbus was created in 1967 with the purpose of strengthening co-operation in developing aviation technology within Europe, in order to be able to compete with the United States, which dominated the aircraft manufacturing sector. Germany, France and Great Britain agreed to collaborate the overpowering market presence of the US-American companies, namely Boeing and the recently merged aircraft manufacturer, McDonnell Douglas. In 1969 the world's first twin-engine wide body passenger aircraft was commissioned, and its design and manufacture became a joint European initiative which has carried on producing aircraft ever since (Airbus 2016a).

4.3 Open Innovation at Airbus

Open innovation is facilitated by an Airbus' global network of accelerator units, called BizLabs. They develop the ideas further that come from start-up programs in other industry sectors, employing collaborative arrangements that Airbus entered into with Microsoft, Google and Orange Fab. The purpose of BizLabs is to transform the latest, most relevant innovative ideas into commercial propositions more rapidly, in order to remain ahead of industry competitors. Small and medium-sized enterprises (SMEs) in the supply chain will receive support by especially created offices in major emerging economies, for instance, the BRIC countries China, India and Brazil.

One focus of Airbus is the protection of its data, since data is considered raw material that represents key future business opportunities. In order to generate more robust data sharing models, as well as new business models, Airbus co-operates with universities and start-ups. In 2015, the A3 Innovation Center was established as a joint venture to quickly exploit innovative and disruptive technologies that will transform aircraft manufacture and operational performance.

The Airbus Group has also initiated a Leadership University opened in 2015 in France, which intends to train future leaders in networking, in order to share knowledge and to generate creative ideas and a culture of collaboration and learning. The University will operate internationally at the main Airbus sites, starting in 2016. In focus of the syllabus are the training of individual skills and personal awareness for employees. University and training center offer 360° feedback courses, mentoring initiatives, and team programs with entrepreneurs and start-up companies involved in leadership development. Leaders will be encouraged to develop partnerships with other companies and universities, and to continually improve business practice and knowledge/trends of the sector (Airbus 2016b).

The Global Engineering Deans' Council (GEDC), which is a global body that recognizes excellence in engineering, is another initiative by Airbus to identify innovative ideas. It has fostered the creation of the Airbus GEDC Prize; this award was introduced for ideas generated by young engineers. The prize money is US $10,000 for the winner and US$1500 for two runner-ups (HRMID 2015).

Jean Botti, the former CTO of Airbus who left the company for Phillips in April 2016, however, was somewhat critical of the company's approach to open innovation in his last speech in his capacity as the company's CTO. "[For inspiration] we look a lot inside, and a lot outside [the company], especially when we consider our 265 partnerships in the academia and national laboratories we manage", he said (Stojanovic 2016). The company would need to realign with the idea of open innovation and to use larger budgets if it wanted to respond effectively to the emerging competition, he stated. The fixed budget for open innovation was too low and resulted in being spread too thinly. "We need to look at the potential of open innovation, dedicating larger budgets to initiatives," Botti urged.

4.4 Open Innovation at Boeing

Boeing runs several open innovation initiatives, for instance, the ecoDemonstrator Program, which facilitates testing technologies for reducing emission and noise as well as enhancing fuel and operational efficiency. Engineers test these technologies on current Boeing models such as 757 and 787 whilst in operation, in order to speed up development, and instead of using slower simulation software.

Employees are encouraged to generate other ideas to resolve environment issues, which represent challenges to the aviation sector. Employee involvement teams are tasked with finding ways to reduce waste, for example (Boeing 2016b).

Boeing also makes use of supplier innovations such as the winglets, which were developed by Aviation Partners (The Wall Street Journal 2012).

The development of the Boeing 787 was meant as an open innovation project in which Boeing changed its approach in how it designed, built and financed new aircraft by much greater involvement of its supply chain partners (Kotha and Srikanth 2013).

The 787 Dreamliner was to be built from a high proportion of composites (Slayton and Spinardi 2016). Major global partners were provided with highly detailed specifications and performance metrics and were expected to provide Boeing with innovative ideas for the aircraft (Kotha and Srikanth 2013). The partnerships including their tier-1 suppliers are shown in detail in Fig. 4.

This approach was in contrast with the previous process in which Boeing engineers created the design and drawings and then provided the suppliers with specifications for hundreds of different parts. There are many sources, however, that regard the building of the 787 Dreamliner as a botched attempt to employ open

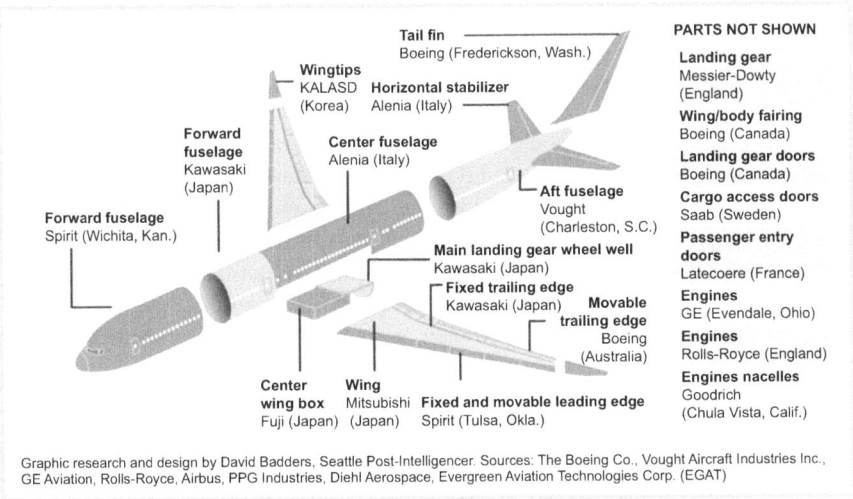

Fig. 4 Global partnerships for Boeing 787. Source: Shenar et al. (2016)

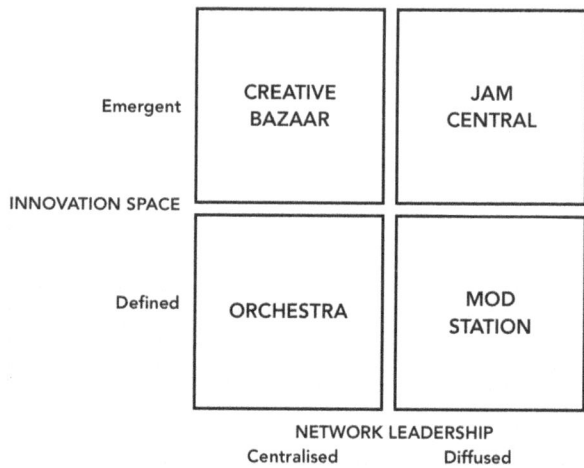

Fig. 5 Open innovation approaches. Source: Sawhney and Nambisan (2007, p. 59)

innovation. Henry Chesbrough, who coined the term, wrote in 2011: "Boeing is a good recent example of the problem of interdependencies in systems design. [...] It broke the design [of the 787] into major subsystems, and contracted with design partners for those major subsystems. However, when the subcontractors came back with their proffered designs, Boeing couldn't get all the pieces to work together. Between the new materials, the new engines, the new avionics, etc., there were too many new and poorly understood aspects of the design to make a working aircraft." (Chesbrough 2011) This does not necessarily imply a failure in Boeing's approach but defines the limits of open innovation.

Sawhney and Nambisan (2007) suggested that there are different approaches to the open innovation concept. Depending on how closely the creative process is controlled and how narrowly the expectations are defined, four archetypes emerge that describe how open innovation can be approached. The archetypes, which take their name from more or less loosely arranged musical groups, are shown below (Fig. 5 is above).

The approach used by Boeing, a company that had relied on in-house R&D for decades, was defined and centralized (Orchestra). However, the high complexity of an aircraft suggests that many developers cooperate on an equal basis, acting as peers. This suggests another approach to open innovation, though. "In sum, the Jam Central model is characterized by shared exploration of an innovation arena by a peer group of contributors who share in the responsibility of directing and coordinating the innovation effort", Sawhney and Nambisan wrote (Sawhney and Nambisan 2007, p. 65).

5 Conclusion

When Airbus was established, it was based on the principles of collaboration with the intention of generating more ideas and better decision making, in order to be competitive (Almirall and Casadesus-Masanell 2010). The evidence demonstrates that it has employed the open innovation model extensively, using the outside-in model, exemplified by creating joint ventures, working with suppliers and gaining external knowledge, for instance, through its suppliers, innovative firms in other sectors, and through its own university (Gassmann and Enkel 2004; DIUS 2008). Airbus has extended open innovation by founding an innovation excellence center in the Silicon Valley (Gassmann et al. 2010).

Boeing made successfully use of open innovation practices in the past, for example, when it adopted the winglet design, which was a supplier innovation, several years before Airbus introduced its Sharklets (The Wall Street Journal 2012). More recently, Boeing adopted open innovation in an attempt to regain competitiveness, but the attempt did not provide the intended results. Academic papers on the development of the 787 Dreamliner point out that Boeing's management underestimated the complexity of innovation. However, the complexity only became obvious through Boeing's trial and error experience. Research that was conducted after Chesbrough published his work on open innovation, which included Boeing as a negative example, suggests that the complexity must be dealt with in different ways.

Simple projects can be controlled by applying a defined and centralized approach, whereas highly complex matters need more than just one developer in control. The "Jam Central" model of a peer group that develops ideas in an attempt to enable open innovation seems to be quite promising and might have been useful in the case of Boeing.

References

Airbus (2016a) A model of discovery and divergence. Early days (1967–1969). Acad Manag Rev. Airbus (online). Available via http://www.airbus.com/company/history/the-narrative/early-days-1967-1969. Accessed 24 May 2016

Airbus (2016b) Flying ahead: airbus annual report 2015. Airbus

Almirall E, Casadesus-Masanell R (2010) Open versus closed innovation: a model of discovery and divergence. Acad Manage Rev 35(1):27–47

GE Aviation (2015) Press release: finding more than 50 new ways to approach inspection. 14 April 2015. Online. http://www.geaviation.com/press/services/services_20150414.html. Retrieved 20 Aug 2016

Baregheh A, Rowley J, Sambrook S (2009) Towards a multidisciplinary definition of innovation. Manag Decis 47(8):1323–1339

Boeing (2016a) A century in the sky. Boeing (online). Available via http://www.theatlantic.com/sponsored/boeing-2015/a-century-in-the-sky/652/. Accessed 24 May 2016

Boeing (2016b) The Boeing Company: 2015 environment report. Boeing

CG (2011) The changing face of the Aerospace and Defence Industry. Capgemini, Paris

Chesbrough H (2003a) Open innovation: the new imperative for creating and profiting from technology. Harvard Business School Press, Boston, MA
Chesbrough H (2003b) The era of open innovation. Sloan Manage Rev 44(3, Spring):35–41
Chesbrough H (2011) Open innovation and the design of innovation work. Forbes, 18 May 2011. Online. http://www.forbes.com/sites/henrychesbrough/2011/05/18/open-innovation-and-the-design-of-innovation-work/#2f5d81407d74
Chesbrough H, West J, Vanhaverbeke W (2006) Open innovation: researching a new paradigm. Oxford University Press, Oxford
Crouch TD (2003) The Bishop's boys: a life of Wilbur and Orville Wright. W. W. Norton & Company, New York
Deloitte (2015) 2015 global aerospace and defense sector financial performance study. Deloitte
Di Minin A, Frattini F, Piccaluga A (2010) Fiat: open innovation in a downturn (1993–2003). Calif Manage Rev 52(3):132–161
DIUS (2008) Innovation nation. Department for Innovation, Universities & Skills, London
Gassmann O, Enkel E (2004) Towards a theory of open innovation: three core process archetypes. R&D Manag Conf 6:1–18
Gassmann O, Enkel E, Chesbrough H (2010) The uture of open innovation. R&D Manag 40(3):1–9
Habicht H, Möslein KM, Reichwald R (2011) Open innovation: Grundlagen, Werkzeuge, Kompetenzentwicklung. Inf Manag Consult 26(1):44–51
Hilgers D, Piller F (2009) Controlling für open innovation, Theoretische Grundlagen und praktische Konsequenzen. Controlling 21(2):5–11
HRMID (2015) Diversity in engineering takes off with Airbus award. Hum Resource Manag Int Dig 23(4):16–17
Ili S, Albers A, Miller S (2010) Open innovation in the automotive industry. R&D Manag 40(3):246–256
Johnson G, Whittington R, Scholes K, Agwen D, Regner P (2013) Exploring strategy text & cases, 10th edn. Pearson, Harlow
Knudsen P, Mortensen TB (2011) Some immediate—but negative—effects of openness on product development performance. Technovation 31:54–64
Kotha S, Srikanth K (2013) Managing a global partnership model: lessons from the Boeing 787 'Dreamliner' program. Glob Strateg J 3:41–66
KPMG (2013) The future of civil aerospace. KPMG, UK
Lewandowska J (2012) Frost & Sullivan: private equity and venture capital investments in aerospace and defence slow due to economic uncertainty in Europe. PR Newswire, 28 June
Lindegaard S (2010) The open innovation revolution: essentials, roadblocks, and leadership skills. Wiley, Hoboken, NJ
Meyer J (2006) What is good qualitative research? A first step towards a comprehensive approach to judging rigour/quality. J Health Psychol 11(5):799–808
OW (2015) Challenges for European aerospace suppliers. Oliver Wyman
PRN (2015) 2015 global aerospace and defense industry to grow around 3 percent. PR Newswire, 12 February
PWC (2012) Sectoral structure analysis aerospace review. Price Waterhouse Cooper, Ontario
Reichwald R, Piller F (2009) Interaktive Wertschöpfung: open innovation, Individualisierung und neue Formen der Arbeitsteilung, 2. Aufl. Wiesbaden
Saunders M, Lewis P, Thornhill A (2016) Research methods for business students, 7th edn. Pearson, Harlow
Sawhney M, Nambisan S (2007) The global brain: your roadmap for innovating faster and smarter in a networked world. Pearson Prentice Hall, Milan
Shenar AJ, Holzmann V, Melamed B, Zhao Y (2016) The challenges of highly complex projects: what can we learn from Boeing's Dreamliner experience? Proj Manag J 47(2):62–78
Slayton R, Spinardi G (2016) Radical innovation in scaling up: Boeing's Dreamliner and the challenge of socio-technical transitions. Technovation 47:47–58

Stojanovic P (2016) Airbus Group's departing CTO on his open innovation strategy. Hottopics, March 2016. Online. https://www.hottopics.ht/stories/enterprise/innovation-inspiration-airbus-group/. Retrieved 28 Aug 2016

The Wall Street Journal (2012) Air war: 'Winglet' versus 'Sharklet'. 6 March 2012. Online. http://www.wsj.com/articles/SB10001424052970204778604577239583270202816. Retrieved 20 Aug 2016

Johannes Walther is chairman of the executive board of the IPM AG - Institute for Production Management and Professor of Business Administration with a focus on production management at the Faculty of Economics at the Ostfalia University of Applied Sciences. He is the publisher of the magazine Supply Chain Management™, Member of the Advisory Board of the Volkswagen Group Institute for Procurement (IFB) and the Academic Forum of Volkswagen Group Procurement, as well as lecturer at the Volkswagen Group AutoUni. He studied economics at the University of Hanover. After earning his doctorate degree in 1992, he assumed leading positions in industry, trading and service companies. Born in Hannover in 1959, Johannes is married with two children and currently lives in Hanover, Germany.

Daniel Wäldchen is member of the executive board at the IPM AG - Institute for Production Management and Ph.D. student at the Bundeswehr University Munich, where he conducts research on Exclusive Innovation Exchange and Supplier Innovation. He serves as a visiting lecturer at the Ostfalia University of Applied Sciences as well as the Volkswagen Group AutoUni and is member of the editorial board of the magazine Supply Chain Management™. Daniel Wäldchen studied economics in Wolfsburg and Berlin and was employed at Volkswagen AG before he joined IPM. Born in Salzgitter in 1985, he currently lives in Berlin, Germany.

Disruptive Innovation Through 3D Printing

Reinhart Poprawe, Christian Hinke, Wilhelm Meiners, Johannes Schrage, Sebastian Bremen, Jeroen Risse, and Simon Merkt

Abstract Emerging technologies such as 3D Printing or Additive Manufacturing (AM) and especially Selective Laser Melting (SLM) provide great potential for solving the dilemma between scale and scope, i.e. manufacturing products at mass production costs with a maximum fit to customer needs or functional requirements. Because of the technology's intrinsic advantages such as one-piece-flow capability and almost infinite freedom of design, Additive Manufacturing was recently described as "the manufacturing technology that will change the world". Due to the complex nature of production systems, the technological potential of AM and particularly SLM can only be realized by a holistic comprehension of the complete value creation chain, especially the interdependency between products and production processes. Therefore, this chapter aims to give an overview on recent research in machine concepts and component design, which experts of the Cluster of Excellence "Integrative production technology for high wage countries" carried out.

1 Additive Manufacturing

Due to technology intrinsic advantages like one-piece-flow capability and almost infinite freedom of design, 3D Printing or Additive Manufacturing (AM) was described as "the manufacturing technology that will change the world" (Economist 2011) and several international research groups are working on this topic (Gibson et al. 2010; Hopkinson et al. 2005; Lindemann 2006).

Research at Fraunhofer Institute for Laser Technology (ILT) and RWTH Aachen University Chair for Laser Technology (LLT) in the area of Additive Manufacturing (AM) is dedicated to fundamental research and applied research, aiming to bridge the gap to industrial production. SLM is a powder bed based process and works without tools; only the digital design of a component is required, resulting in a great freedom of geometry, shown exemplarily in a monolithic turbine seal with honeycomb (see Fig. 1, wall thickness of honeycombs less than 100–200 µm).

R. Poprawe • C. Hinke (✉) • W. Meiners • J. Schrage • S. Bremen • J. Risse • S. Merkt
Fraunhofer Institute for Laser Technology, RWTH Aachen University, Aachen, Germany
e-mail: christian.hinke@ilt.fraunhofer.de

Fig. 1 Monolithic turbine seal with honeycomb structures, manufactured by SLM in cobalt based super alloy

One overall objective of the Aachen based Cluster of Excellence "Integrative production technology for high wage countries" is the resolution of the dichotomy between scale and scope, i.e. manufacturing products at mass production costs with a maximum fit to customer needs or functional requirements (Schleifenbaum et al. 2011).

The emerging Additive Manufacturing (AM) and especially the Selective Laser Melting (SLM) technologies provide great potential for solving this dilemma. With this layer-based technology, the most complex products can be manufactured without tools. The 3D CAD model gets sliced layer wise for computing the scan tracks of the laser beam. In a first manufacturing step, powder material (typically in a range of 25–50 μm) is deposited as a thin layer (typically 30–50 μm) on a substrate plate. According to the computed scan tracks, the laser beam melts the powder, which is solidified after melting. The substrate plate is lowered and another powder layer is deposited onto the last layer and the powder is melted again to represent the parts geometry (see Fig. 2). These steps are repeated until almost 100 % dense parts with serial-identical properties are manufactured with the SLM process directly from the 3D CAD model (Meiners 1999; Over 2003; Schleifenbaum et al. 2008).

AM technology in general and SLM in particular is characterized by a fundamentally different relation of cost, lot size and product complexity compared to conventional manufacturing processes (see Fig. 3). There is no increase of costs for small lot sizes (in contrast to mold-based technologies) and no increase of costs for shape complexity (in contrast to subtractive technologies).

For conventional manufacturing technologies such as die casting, the piece cost depends on the lot size. For increasing lot sizes, the piece costs are decreasing due to economies-of-scale. Because lot dependent fixed costs (e.g. tooling costs) are very low for AM, AM enables economic part production in small lot sizes (Individualization for free). Innovative business models such as customer co-creation can be implemented, using this advantage of AM technologies (see Fig. 3 left).

The more complex a product is, the higher the piece cost for manufacturing. This relation does not apply for AM. The nearly unlimited geometric freedom that AM offers makes the piece cost almost independent from product complexity. In some

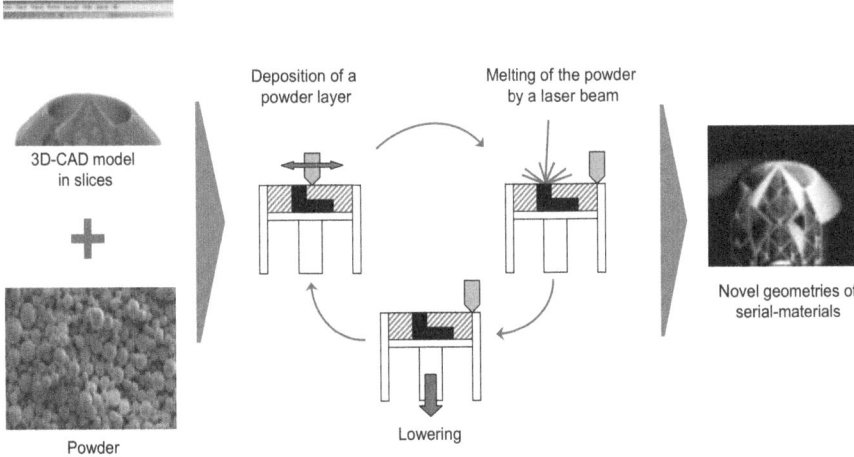

Fig. 2 Schematic representation of the SLM process

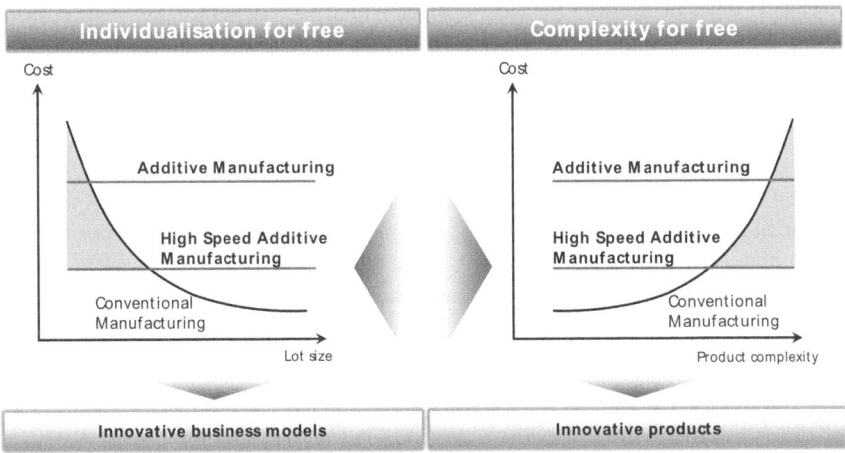

Fig. 3 Innovative business models and innovative products enabled by Additive Manufacturing

cases, manufacturing costs can even decrease due to lower build-up volumes of optimized products with high geometric complexity. Topology optimization is one design approach to save weight while functionally adapting the product design to predefined load cases (Huang and Xie 2010). These different relations between piece cost and product complexity offer a unique capability for AM to manufacture innovative products, perfectly adapted to the specific technological requirements through the integration of lattice structures (see Fig. 3 right).

2 Additive Manufacturing in Turbomachinery

Companies in the field of turbomachinery have started to use AM for manufacturing prototypes or parts for product development and rig tests to profit, for instance, from shorter lead times and therefore faster design-test iterations or an easier integration of measuring equipment (Economist 2011). In these cases, disadvantages such as high manufacturing costs, low productivity or not sufficient mechanical properties, because of unavailable materials, are acceptable. Rig-parts, for instance, have to withstand the test conditions only for a limited amount of time, so that alloys with lower mechanical properties can be used, instead of alloys of series parts.

Meanwhile, several major OEMs have announced to start series production of components with SLM for aero engines. Two of the most known components are the fuel nozzle out of a cobalt-chrome alloy (Rockstroh et al. 2013), and the borescope boss out of the nickel based super alloy Inconel® 718 (IN718) (MTU 2013). Parts like borescope bosses are typically casted or milled, leading to high production costs due to difficult machining of these materials. Near-net shape manufacturing achieved by SLM can lead to a significant reduction of post-processing costs and allows for new designs, which are conventionally not producible (MTU 2013). This freedom of geometry is also the main driver for the fuel nozzle. The design focuses on significantly increasing durability and functionality, whereby reducing the amount of individual parts required to be produced and joined, so the nozzle can be manufactured in nearly one piece (GE 2014).

Despite of the novelty of the production itself, significant differences exist between both components regarding required quantity, type of load and especially risk for an engine breakdown in case of a component failure. The borescope bosses are not flight-critical as they are placed on the engine casing and allow the blades to be inspected for wear and damage. This leads to considerable lower thermal and mechanical loads in comparison to the fuel nozzle mounted in the engine combustor. The conditions in the engines hot path lead to high thermal-cyclic stresses, and the functionality of the combustor depends on the fuel nozzle making it flight-critical. From a production point of view, the major difference is the required quantity of both parts: approx. 1–2 borescope bosses and 19 fuel nozzles are needed per engine. The borescope boss will be produced in medium scale series, whereas the required amount of fuel nozzles is estimated to be >100.000 in the next 4–5 years, implicating a large-scale series production (GE 2014; MTU 2013).

The repair of components such as burners for stationary gas turbines is already carried out in smaller scale, using SLM. The tip area of these complex components is exposed to high temperatures, leading to substantial wear and damage during operation. The conventional repair chain includes several intermediate steps for part removal, joining operations and inspection. In the SLM process chain, first the damaged tip is completely removed. In a second step, the tip is built directly on top of the remaining part of the component, yielding a shorter lead-time and reduced costs due to less process steps. Additionally, less obvious benefits arise because of this repair method, for example, the possibility to introduce an improved design of the burner tip directly, to enhance the fuel efficiency in older turbine models (EOS 2014).

The described examples for series production or repair show that the SLM technology becomes mature and begins to establish itself as an additional production technology in turbomachinery. Key drivers for this are the technology's design freedom and great flexibility during production. Nevertheless, these drivers need a robust and comprehensive technical basis to achieve the required part qualities and the material properties in combination with a high productivity and a suitable quality management. Applications such as flight-critical components in aero-engines or key-components in stationary gas turbines show that this basis exists, but also needs improvement as the low productivity of the current machine technology, for example, is not sufficient.

Because of the complex nature of production systems, the technological potential of AM and particularly SLM can only be realized for turbomachinery by a holistic comprehension of the complete value creation chain, especially the interdependency between products and production processes.

Therefore, this article aims to give an overview regarding recent research in SLM machine concepts and component design, which has been carried out within the Cluster of Excellence "Integrative production technology for high wage countries" and the research campus "Digital Photonic Production".

3 SLM Machine Concepts

State-of-the-art SLM machines are typically equipped with a 200 or 400 W laser beam source and provide a build space of $250 \times 250 \times 300$ mm^3. There are different ways to increase the productivity of SLM machines in terms of process build rate. Productivity can be generally increased by the following measures: increase of laser power to boost scanning speed, using bigger layer thickness and bigger beam diameter to raise the build-up speed.

Another method to increase productivity is: SLM process parallelization by using multiple laser beam sources, and employing multi laser-scanning-systems in one machine. Either the build area can be multiplied, or one build space can be processed by multi-lasers and scanning-systems at the time.

3.1 High Power Selective Laser Melting

Recent developments in SLM machines show that machine suppliers offer SLM systems with increased laser power ($P_L \leq 1$ kW). The aim is to increase the process speed and thereby the productivity of the SLM process. However, by the use of a beam diameter of approx. 100 µm, commonly used in commercial SLM systems, the intensity at the point of processing is significantly increased due to the use of increased laser power. This effect results in spattering and evaporation of material and therefore in an unstable and not reproducible SLM process. For this reason, the beam diameter has to be increased in order to lower the intensity in the processing zone. In this case, the melt pool is increased and the surface roughness of the manufactured part is negatively influenced. To avoid these problems, the so-called skin-core strategy (see Fig. 4) is used, here the part is divided into an inner core and an outer skin (Schleifenbaum et al. 2010). Different process parameters and focus diameters can be designated to each area. The core does not have strict limitations or requirements concerning the accuracy and detail resolution. Therefore, the core area can be processed with an increased beam diameter (ds = 400–1000 µm) and an increased laser power; thus resulting in an increased productivity. On the other hand, the skin area is manufactured with the small beam diameter (ds = 70–200 µm) in order to assure the accuracy and surface quality of the part.

When increasing the beam diameter and adapting the process parameters, the cooling and solidification conditions change significantly in comparison to the conventional SLM process. Therefore, the microstructure and the resulting mechanical properties have to be investigated in detail. The Cluster of Excellence investigated the maraging tool steel 1.2709.

The first step is to investigate process parameters on cubic test samples which have an averaged density of $\geq 99.5\,\%$. Therefore, an SLM machine setup with a laser beam diameter of ds = 80 µm (Gaussian beam profile) and ds = 728 µm (Top-hat beam profile) is used. The results for the achieved theoretical build-up rate are calculated by adding the product of hatch distance, layer thickness and scanning velocity. It can be observed that by the increase of the laser power from 300 W up to $PL = 1$ kW and an adaption of the process parameters layer thickness and scanning velocity, the theoretical build up rate can be increased from 3 to 15 mm³/s. A further increase of the laser power up to $PL = 2$ kW results in an increase of the theoretical build-up rate to 26 mm³/s (factor 8,9). These investigations show that it is possible to heighten the theoretical build-up rate and the productivity significantly by using increased laser power up to $PL = 2KW$.

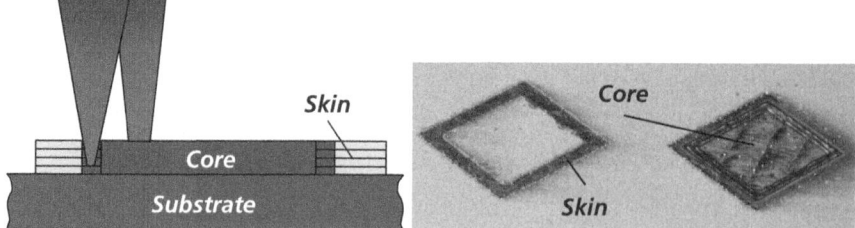

Fig. 4 Principle skin-core strategy

3.2 Multi-Scanner Selective Laser Melting

Experts of the Cluster of Excellence designed and implemented a multi-scanner-SLM machine. Two lasers and two scanners are integrated in this system. These two scanners can be positioned to each other, either that both scan two fields on its own (double-sized build space) or that one scan field is processed with two scanners simultaneously (see Fig. 5).

New scanning strategies can be developed and implemented with these multi-scanner systems (see Fig. 6).

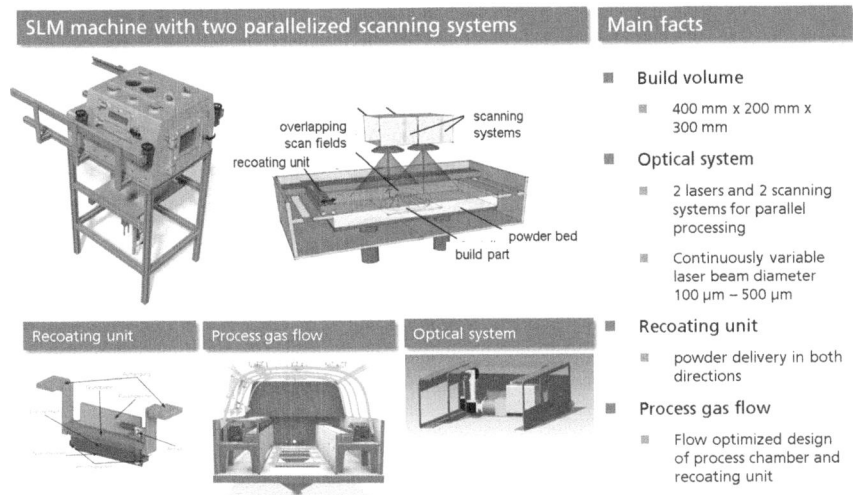

Fig. 5 SLM machine concept of parallelization

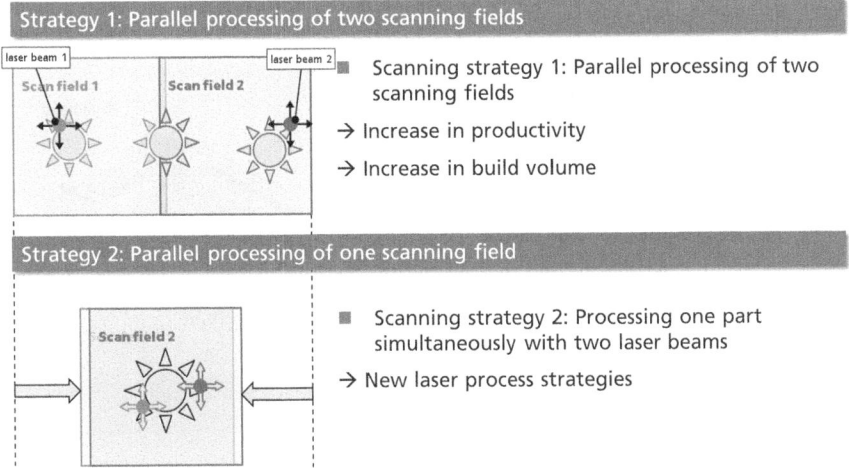

Fig. 6 New SLM laser processing strategies with two lasers and two scanning systems

Scanning strategy 1: both scan fields are positioned next to each other with a slight overlap. This results in a doubling of the build area. By using two laser beam sources and two laser-scanning systems, both scan fields can be processed at the same time. In this case, the build-up rate is doubled.

Scanning strategy 2: the two laser beam sources and the two laser-scanning systems expose the same build area. Again, a doubling of the build-up rate is achieved. In addition, new process strategies may be developed: a laser beam is used for preheating the powder material that is followed by a second laser beam that melts the powder afterwards.

3.3 Multi-Spot Selective Laser Melting

Another way to increase productivity is by parallelization of the SLM process using multiple laser beam sources and multi laser-scanning systems in one machine (see Fig. 7).

The authors have developed a new way, based on multi-diode lasers. The machine design uses no scanner system, and relies on a printer head instead, featuring several individually controllable diode lasers that move using linear axes.

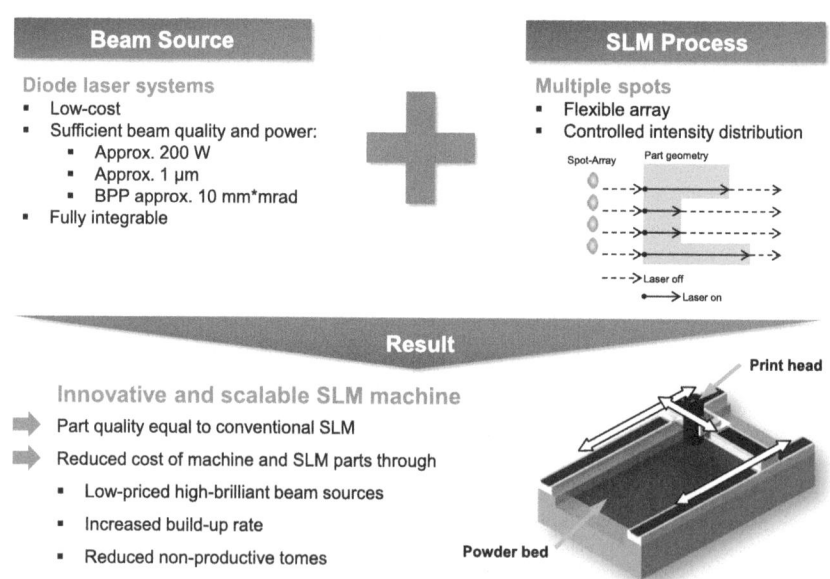

Fig. 7 New multi-spot SLM system based on diode lasers

The advantage of multi-spot processing: the system's build-up rate can be increased by adding a virtually unlimited number of beam sources—with no need for modifications to the system design, to the exposure control software or to process parameters.

The new machine design allows to increase building space, simply by extending the travel lengths of the axis system, without changing the optical system. In addition, the processing head has a local shielding gas flow system that guarantees a constant stream of shielding gas at each processing point, regardless of the size of the installation space. This is essential for achieving position-independent, reproducible component quality.

4 Functional Adapted Component Design

As explained in Fig. 2, Additive Manufacturing in general and SLM in particular provides a great potential for innovative business models and innovative products or components. Due to technology intrinsic advantages such as "Individualization for free" and "Complexity for free", SLM is the technology of choice for the production of functional adapted products or components in small or medium lot sizes.

In order to raise this potential, design methods have to be adapted to the potential of Additive Manufacturing (such as topology optimization) or even new design methods have to be developed (such as lattice structures). The following part shows recent results in the field of topology optimization and lattice structure design.

4.1 Topology Optimization and SLM

In contrast to subtractive manufacturing methods like machining, the main cost driver of the SLM process is the process time, needed to generate a certain amount of part volume. In consequence, reducing the part volume to the lowest amount needed to absorb the forces of the use case, is an important factor to increase the productivity of the SLM process. Topology optimization is an instrument to design load adapted parts based on a FEM-analyzes. The load cases, including forces and clamping, need to be very clear to get the best results possible. In an iterative process, the topology optimization algorithm calculates the stress level of each FEM element. Elements with low stresses are deleted until the optimization objective/criterion (such as weight fraction) is reached. The topology optimization usually results in very complex parts with 3D freeform surfaces, hollow and filigree structures. The commonly used way to fabricate these part designs is a reconstruction considering process restrictions of conventional manufacturing methods such as casting or machining, resulting in a lower weight reduction. SLM opens opportunities to fabricate complex optimization designs without any adjustments after optimization.

The reduction of weight is an important factor in aerospace industry (Rehme 2009). The weight of the aircraft determines fuel consumption. An aircraft seat manufacturer is investigating the opportunities to save weight in their business class seats through SLM. One part of the seat assembly, a kinematics lever, was selected to investigate the potential of the direct fabrication of topology optimization results via SLM (see Fig. 8).

In a first step, the maximum design space and connecting interfaces to other parts in the assembly were defined (see Fig. 8) to guarantee the fit of the optimization result to the seat assembly. Interfacing regions are determined as frozen regions, which are not part of the design space for optimization. The kinematics lever is dynamically loaded if the passenger takes the sleeping position. Current topology optimization software is limited to static load cases. Therefore, the dynamic load case is simplified to five static load cases, which consider the maximum forces at different times of the dynamic seat movement. Material input for the optimization is based on an aluminum alloy (7075) which is commonly used in aerospace industry: material density: 2810 kg/m^3, E Modulus: 70.000 MPa, Yield Strength: 410 MPa, Ultimate Tensile Strength: 583 MPa and Poisson's Ratio: 0.33. The objective criterion of the optimization is a volume fraction of 15 % of the design space. The part is optimized regarding stiffness. Figure 9 shows the optimization result as a mesh structure and an FEM analyzes for verification of the structure.

The maximum stress is approx. 300 MPa, which is below the limit of Yield Strength of 410 MPa. Before manufacturing, the surfaces of the optimized part are smoothened to improve the optical appearance of the part. Compared to a conventional part (90 g), a weight reduction of approx. 15 % (final weight 77 g, see Fig. 10) was achieved. Further improvements to increase the productivity of the process are needed for series part production.

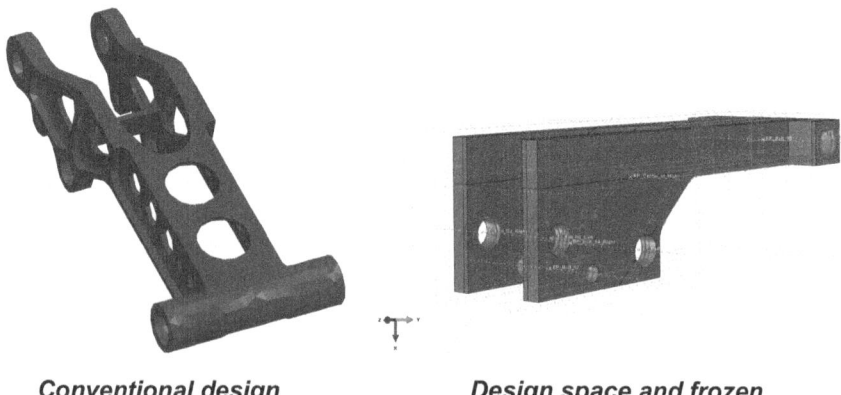

Conventional design **Design space and frozen regions (red)**

Fig. 8 Kinematics lever of a business class seat

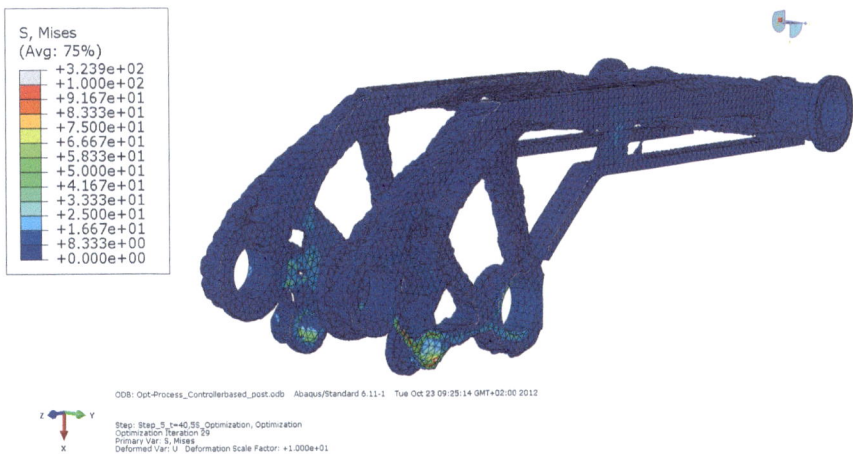

Fig. 9 Mesh structure of optimization result including stress distribution

Fig. 10 Final lightweight part manufactured by SLM

4.2 *Functional Adapted Lattice Structures and SLM*

The almost unlimited design freedom offered by SLM provides new opportunities in lightweight design through lattice structures. Due to unique properties of lattice structures (good stiffness to weight ratio, great energy absorption, etc.) and their low volume, the integration of functional adapted lattice structures in functional parts is a promising approach for using the full technology potential of SLM (see Fig. 11). Compared to conventional manufacturing technologies, piece costs of SLM parts are independent of part complexity and the main cost driver is the process time (correlates with part volume). Lattice structures can reduce the amount of part volume and host unique properties.

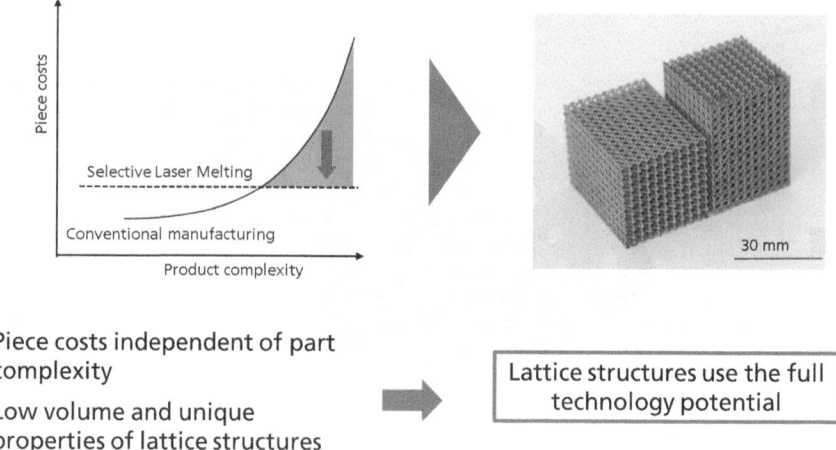

Fig. 11 Complexity-for-free offers great opportunities through lattice structures

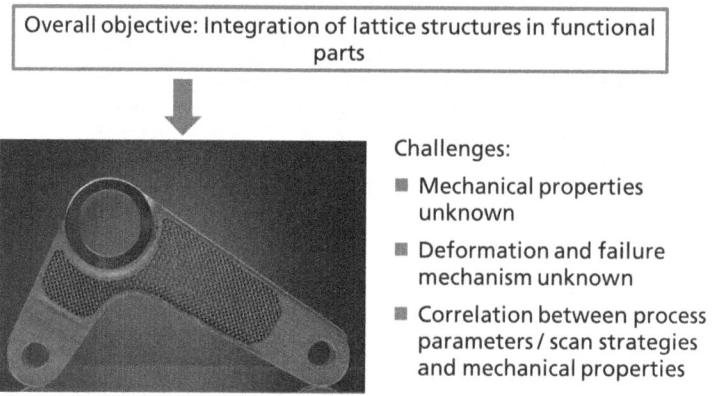

Fig. 12 A new way of designing functional parts by the integration of lattice structures

Three main challenges need to be solved to make lattice structures a real option for the use in functional parts in different industries. Several researchers studied the mechanical properties of different lattice structure types (Löber et al. 2011; Shen et al. 2012; Yan et al. 2012; Rehme 2009; Gümrük and Mines 2013; Smith et al. 2013; Ushijima et al. 2011; Shen et al. 2010). Nevertheless, there is no comprehensive collection of mechanical properties of lattice structures under compressive, tensile, shear and dynamic load. The deformation and failure mechanisms are not studied sufficiently. A relatively new field of research is the influence of different scan parameters/strategies on mechanical properties. To reach the overall objective of our research, these challenges need to be overcome to design functional adapted parts with integrated lattice structures (see Fig. 12).

References

Economist (2011) Cover story. 2 Dec 2011

EOS GmbH (2014) Industry: Siemens—EOS Technology opens up new opportunities for industrial gas turbine maintenance cost reduction. Available via http://www.eos.info/press/customer_case_studies/siemens. Accessed 23 Jun 2016

Gibson I, Rosen D, Stucker B (2010) Additive manufacturing technologies. Rapid prototyping to direct digital manufacturing. Springer, Heidelberg

GE Report (2014) Fit to print: new plant will assemble world's first passenger jet engine with 3D printed fuel nozzles, next-gen materials. Available via http://www.gereports.com/post/80701924024/fit-to-print/. Accessed 23 Jun 2016

Gümrük R, Mines RAW (2013) Compressive behaviour of stainless steel micro-lattice structures. Int J Mech Sci 68:125–139

Hopkinson N, Hague R, Dickens P (2005) Rapid manufacturing: an industrial revolution for the digital age. Wiley, Chichester

Huang X, Xie M (2010) Evolutionary topology optimization of continuum structures: methods and applications. Wiley, Chichester

Lindemann U (2006) Individualisierte Produkte—Komplexität beherrschen in Entwicklung und Produktion. Springer, Heidelberg

Löber L, Klemm D, Kühn U, Eckert J (2011) Rapid manufacturing of cellular structures of steel or titaniumalumide. Mater Sci Forum 690:103–106

Meiners W (1999) Direktes Selektives Lasersintern einkomponentiger metallischer Werkstoffe. Dissertation. RWTH Aachen

MTU Aero Engines (2013) Trailblazing technology: MTU Aero Engines produces parts by additive manufacturing. Available via www.mtu.de/fileadmin/EN/7_News_Media/1.../Additive_Manufacturing.docx. Accessed 23 Jun 2016

Over C (2003) Generative Fertigung von Bauteilen aus Werkzeugstahl X38CrMoV5-1 und Titan TiAL6V4 mit "Selective Laser Melting". Dissertation. RWTH Aachen

Rehme O (2009) Cellular design for laser freeform fabrication. Dissertation. Laser Zentrum Nord, Hamburg

Rockstroh T, Abbott D, Hix K, Mook J (2013) Additive manufacturing at GE Aviation, Industrial Laser Solutions. Available via http://www.industrial-lasers.com/articles/print/volume-28/issue-6/features/additive-manufacturing-at-ge-aviation.html. Accessed 23 Jun 2016

Schleifenbaum H, Meiners W, Wissenbach K (2008) Towards rapid manufacturing for series production: an ongoing process report on increasing the build rate of Selective Laser Melting (SLM). International conference on rapid prototyping & rapid tooling & rapid manufacturing

Schleifenbaum H, Meiners W, Wissenbach K, Hinke C (2010) Individualized production by means of high power selective laser melting. CIRP J Manuf Sci Technol 2(3):161–169

Schleifenbaum H et al (2011) Werkzeuglose Produktionstechnologien für individualisierte Produkte. In: Brecher C (ed) Integrative Produktionstechnik für Hochlohnländer. Springer, Heidelberg, pp 143–181

Shen Y, McKown S, Tsopanos S, Sutcliffe CJ, Mines RAW, Cantwell WJ (2010) The Mechanical Properties of Sandwich Structures Based on Metal Lattice Architectures. J Sandw Struct Mater 12:2010

Shen Y, Cantwell WJ, Mines RAW, Ushijima K (2012) The properties of lattice structures manufactured using selective laser melting. AMR 445:386–391

Smith M, Guan Z, Cantwell WJ (2013) Finite element modelling of the compressive response of lattice structures manufactured using the selective laser melting technique. Int J Mech Sci 67:28–41

Ushijima K, Cantwell WJ, Mines RAW, Tsopanos S, Smith M (2011) An investigation into the compressive properties of stainless steel micro-lattice structures. J Sandw Struct Mater 13(3):303–329

Yan C, Hao L, Hussein A, Raymont D (2012) Evaluations of cellular lattice structures manufactured using selective laser melting. Int J Mach Tools Manuf 62:32–38

Prof. Dr. Reinhart Poprawe holds an M.A. degree in physics from the California State University in Fresno, which he received in 1977. After completion of his diploma and Ph.D. in physics (Darmstadt 1984), he joined the Fraunhofer Institute for Laser Technology in Aachen in 1985, where he began working as Head of the Department "Laser oriented process development". From 1989 to 1/1996 he was Managing Director of Thyssen Laser Technik GmbH in Aachen. Since February 1996, he is Managing Director of the Fraunhofer Institute for Laser Technology and holds the University Chair for Laser Technology at the RWTH-Aachen University. Currently, he is a member of the board in the AKL Arbeitskreis Lasertechnik e. V. Aachen. In 1998, Prof. Poprawe was elected to Fellow in the Society of Manufacturing Engineers in USA (SME). In 2006, he became Fellow of the Laser Institute of America LIA, and in 2012 he became Fellow of SPIE. Since 2001, he is a member of the board of the Laser Institute of America (LIA) and serves on many national and international boards as advisor, referee or consultant, for instance, at the National Laser Centre of South Africa NLC. Over the period of 09/2005–09/2008, he was Vice-Rector for Structure, Research and Junior Academic Staff. He still is chairing the RWTH-International Board, he is the Rectors delegate for China, he received an Honorary Professorship from the Tsinghua University, and he received the Schawlow Award of the Laser Institute of America in 2014.

Christian Hinke studied physics with focus on laser technology at the RWTH-Aachen University. He obtained his diploma degree in 1998. He has been employed at the Chair for Laser Technology (LLT) at RWTH-Aachen University as a scientist since then, and is actually working in the field of "Design for Additive Manufacturing" leading the specialized design group.

Wilhelm Meiners is physicist with a Ph.D. degree in mechanical engineering. He has been employed at the Fraunhofer ILT as a scientist since 1995. In this period, he has developed the Additive Manufacturing process Selective Laser Melting (SLM) and has conducted several national and international research projects. Dr. Meiners is responsible for the Rapid Manufacturing group at Fraunhofer ILT which focuses on SLM.

Johannes Schrage studied mechanical engineering at the RWTH Aachen University and has specialized in production technologies. He obtained his diploma degree in 2011 and has pursued his Ph.D. in the field of Additive Manufacturing at the Chair for Laser Technology at RWTH Aachen University, since then. His main research focus is the machine development for SLM.

Sebastian Bremen studied industrial engineering with focus on production engineering at the RWTH Aachen University. He obtained his diploma degree in 2011 and has pursued his Ph.D. in the field of Additive Manufacturing at the Fraunhofer ILT in Aachen, since then. His main research focus is to increase the productivity and efficiency of the SLM process. He is leading a team focusing on enhancing the productivity of SLM since 2015.

Jeroen Risse studied mechanical engineering at the RWTH Aachen University and has specialized in production technologies and material science. He obtained his diploma degree in 2011 and has pursued his Ph.D. in the field of Additive Manufacturing at the Fraunhofer ILT in Aachen, since then. His main research focus is the processing of nickel- and cobalt-based superalloys by SLM. He is leading a team focusing on turbomachinery and SLM, since 2015.

Simon Merkt studied industrial engineering with focus on mechanical engineering at the Friedrich-Alexander-University in Erlangen-Nürnberg. He obtained his diploma degree in 2011 and pursued his Ph.D. in the field of Additive Manufacturing at the Chair for Laser Technology

LLT in Aachen after. In 2015, he finished his Ph.D. studies on the qualification of additive manufactured lattice structures and its mechanical properties. Currently he is working for TRUMPF Laser- und Systemtechnik GmbH in Ditzingen as an application manager for Additive Manufacturing.

Part II
Configuration and Demand

Fulfil Customer Order Process: Customization of Commercial Aircraft

Gabriel Oehme

Abstract The aircraft industry has seen continuous growth over the past 20 years and is forecast to more than double over the next 20 years, representing a market of more than 30,000 new aircraft. The airline industry has become a very competitive market with new entrants challenging incumbents. The aspect of differentiation has become very important and airlines focus strongly on this when defining new aircraft fleets. While aircraft manufacturers are concerned about increasing production rates to meet future aircraft demand, the customization aspect has to be well considered to not become a tripping stone for aircraft deliveries. Already at aircraft design, the customization features must be taken into account to protect against later surprises and to avoid unnecessary supply chain risks. Comparing two major Airbus aircraft models, the A380 and the A350XWB, will give some insights on how customization has been addressed.

1 Introduction

The aircraft industry has seen strong growth over the last years and is heading for a healthy future with significant annual growth rates over the next 20 years, according to the market forecasts published by Airbus and Boeing (Airbus 2016; Boeing 2016). Since the deregulation of the air transport market in 1978 (Aviation Week 2015) and the creation of Open Skies agreements (U.S. Department of State 2016), and more recently the market entry of new carriers, like the gulf carriers Emirates, Qatar Airways and Etihad Airways, the competition for passengers has changed in pace. The airlines' efforts to attract customers through distinct and highly differentiating products have significantly increased. While aircraft manufacturers are raising aircraft delivery rates to meet the rapidly growing demand, they are exposed to an increasing complexity of the cabin products, at the same time, because airlines like to offer unique branding. Recent examples have shown that the issues on

G. Oehme (✉)
Airbus, Blagnac Cedex, France
e-mail: gabriel.oehme@airbus.com

customized cabin products can even result in aircraft delivery delays like it happened to Boeing with American Airlines on B787 and Airbus on the first A350 to Qatar Airways (Garcia 2014). This article will give insights into the aircraft customization and the impact on the supply chain of the aircraft manufacturers, with a more detailed look into two major Airbus aircraft programs.

2 Basics on Aircraft Purchase Decisions

Aircraft manufactures and OEM suppliers are continuously striving to introduce the most modern and highly competitive products, to enable airlines to efficiently deploy their fleet strategy and network capacity. The right aircraft fleet mix across a regional or global network will provide them the platform for revenue growth and profit generation. However, prior to defining the specific requirements for an aircraft or a fleet, airlines are assessing in detail their current operations, like their route network and its envisaged expansion, aircraft load factors and yield, passenger volumes and segmentation (such as business class, economy class, etc.). Once this evaluation is completed, airlines will be able to express their requirements to the aircraft manufacturers, usually in form of a request for proposal. The request for proposal expresses the future aircraft demand combined with a set of key requirements and evaluation criteria, which will drive the decision process of their future fleet acquisition. One of the typical requirements is the capability of an aircraft to efficiently operate a certain network—basically flying distance in accordance with airline operating rules—with a given number of passengers in different seating classes, and of course a certain additional cargo payload. Passengers and cargo capacity will generate revenues or yield (measurement of fare paid per mile), while for most of the commercial airlines the passenger yield is the primary focus. Aircraft operating cost—in simplistic terms—can be summarized in aircraft acquisition cost, or respectively its depreciation, flight and cabin crew, fuel consumption, and maintenance, to mention only the most prominent ones. In short, airlines will seek achieving high yields while keeping operating cost low.

During the selection process for an aircraft fleet acquisition, the evaluation of an aircraft model can include different engine models and aircraft system choices. However, high attention will be given in particular to the passenger cabin definition. For the purpose of this article, only the cabin will be in focus considering specific market mechanisms and related challenges for aircraft manufacturer and suppliers.

The cabin and its capability to seat a maximum passenger number at a given comfort, usually expressed in seat width and seat pitch, will determine the passenger revenue potential of the aircraft. The available floor space of a given aircraft cabin has to be optimized to reach the best seating capacity, taking into account the airlines' cabin operating rules such as minimum aisle space, and other airline standards for amenities like bars, galleys, and lavatories. The most efficient layout will significantly influence the chances of success in airline sales campaigns. Aircraft manufacturer together with their OEM partner work out the proposition.

A typical way of analyzing and comparing aircraft cabin efficiency is by comparing their respective floor plans or Layout of Passenger Accommodations (LOPA). LOPA examples of an Airbus A380 and an A350XWB are shown below, to make it particularly clear (Fig. 1).

LOPAs are being designed by specialists of the aircraft manufacturers marketing team, focusing on optimizing the floor space usage by intelligently positioning and combining the different cabin monuments like galleys, lavatories, and closets—mainly in door zones—and thereafter installing the requested seating capacity in between the door zones. Cabin monuments are customized to unique airline requirements. Galleys, for example, need to match the catering needs for the different seating categories and airline standards. Airlines will typically define the number of trolleys required for catering, the number of standard units (storage containers) for other items linked to passenger servicing, and also the galley inserts, which are the technical equipment such as coffee makers, water heaters, steam ovens, fridges and freezers, to mention some typical examples. Airlines will have particular requirements for other amenities like specific bars for premium passengers, dedicated storage compartments, and certainly the lavatory count and size, which can also vary by seating class.

Designing and optimizing the LOPA is a crucial activity during aircraft sales campaigns, as it directly impact the revenue potential of the proposed aircraft model, inherently linked to the seating capacity in the different seating categories. As introduced above, airline economics are largely depending on seat count, which will drive the airline revenues and yield potential, and which needs to be balanced against the seat mile cost of a given aircraft with a specific LOPA. The importance of the best aircraft layout is reflected in the high number of LOPA variations usually being studied during sales campaigns, easily exceeding 50 LOPAs for a given aircraft model.

Once a sales campaign has been conclusive, the aircraft purchase agreement will include a LOPA, or multiple ones if several fleets are subject to the aircraft purchase agreement. The aircraft purchase agreement covers the purchase of a certain aircraft model according to standard specifications, and provides for an additional budget linked to the aircraft customization through customer specific changes. These

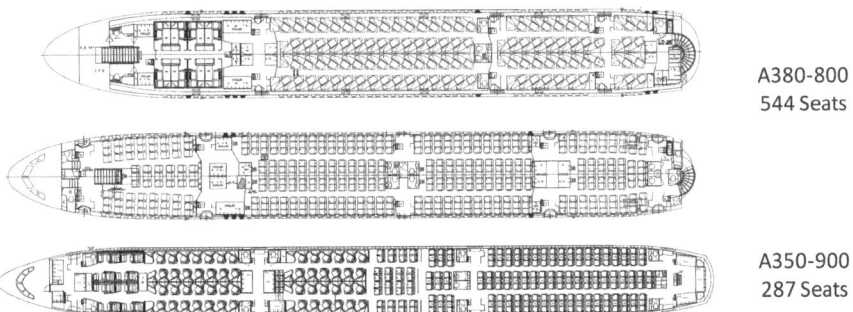

Fig. 1 Marketing LOPA for A380 and A350XWB in four-class layout

changes are mainly related to the customer specific items described in the LOPA. At the time of closing the aircraft purchase agreement, the actual customer aircraft is not yet in production and often several years will lapse until the planned delivery date, mainly due to the significant order backlog at the aircraft manufacturers. This time lapse can induce changes to the initially planned aircraft mission, and can therefore impact its customization in case market conditions have changed.

3 Aircraft Customization Process

At the signature of an aircraft purchase agreement, the aircraft is being sold in accordance with a certain design standard, which reflects basically the technical and the performance baseline for a given aircraft model. This standard is described in the Aircraft Standard Specification, which is structured in accordance with the so called ATA chapters. These are common referencing standard for all commercial aircraft documentation (Air Transport Association of America 1956). Every ATA chapter has a specific scope, from general information like dimensions to technical description of aircraft systems, aircraft structure, cabin interior, power plant, and specific operating procedures. Any modification to this standard will be addressed through customization, which concerns changes to aircraft performance, specific aircraft system additions, and also the definition of the airline specific cabin configuration. Changes to aircraft systems and performance such as take-off or payload capability are usually immediately feasible, because aircraft manufacturers and OEMs develop and certify the needed systems in advance, according to the general market requirements. Cabin customization is rather specific to each airline as it plays a key role in product differentiation in the market. Design, development and certification of the rather unique configurations are carried out for the specific customer version only, a so-called customer head of version. A head of version aircraft is the first aircraft of a customer fleet, all subsequent aircraft are called re-build aircraft.

Aircraft customization follows a well-defined process, the Fulfil Customer Order (FCO) process. The objective of the FCO process is to ensure the delivery of a customized aircraft on-time, on-quality and within a given budget to customer satisfaction. Throughout the entire process, the aircraft manufacturer has installed progressive milestones with respective Quality Gates, to ensure that deliverables are provided in-time and on the required quality. Depending on the aircraft definition and customization progress, different deliverables are required regarding functions such as engineering, manufacturing and procurement, also from external parties, mainly suppliers involved in customization. The FCO process can be structured in four main phases: Definition, Design, Build and Delivery (Fig. 2).

During the Definition Phase, which usually lasts from a few months to up to a year, the aircraft manufacturer's customization team together with the airline specialists define all the required changes to be made to the standard aircraft definition. The starting point for this discussion is the LOPA, which had been

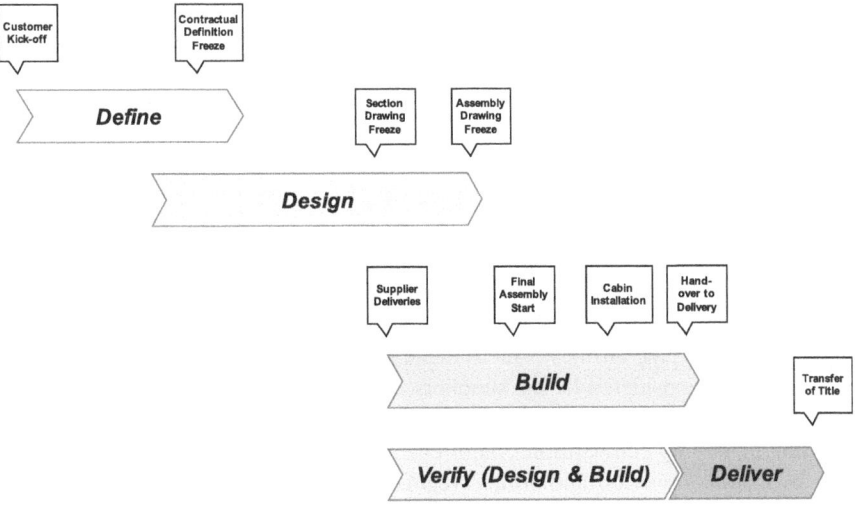

Fig. 2 Fulfil customer order process (Manual M20196; Airbus S.A.S. 2015)

used during the sales campaign and which became part of the aircraft purchase agreement. However, because of the time lapse between signing the aircraft purchase agreement and the planned delivery date, the market requirements might have changed as well as the available cabin options.

Airlines strive for differentiation and the cabin is their main interface to airline customers. This is one of the main reasons, why the cabin is of highest attention within the airline, making the aircraft cabin customization quickly a complex project, especially for long haul aircraft. The cabin has the particularity that the various elements, which are composing an aircraft cabin, are not provided exclusively by the aircraft manufacturer, but rather by a number of competing OEM suppliers. Depending on the aircraft manufacturers' sourcing strategy, airlines can choose certain cabin commodities from multiple suppliers, which each market their products directly to the airlines. This type of supply is called Buyer Furnished Equipment (BFE), as the airline is directly engaging with the supplier into a contractual relationship, while the aircraft manufacturer will ensure installation into the aircraft. The aircraft manufacturer can also decide to source certain equipment directly from the market, and sell it to the airlines as part of the available customization options. This type of supply is called Seller Furnished Equipment (SFE).

At the end of the Definition Phase an agreed aircraft definition, called the Contractual Definition Freeze (CDF), is established with the airline, which then becomes part of the aircraft purchase agreement. The CDF is one of the major milestones in the customization process, as it represents the aircraft version handover (head of version) to the Design Phase. The customization combines BFE and SFE components, and therefore its definition is required to achieve the CDF. The Design Phase is the start of the detailed design activities on both the aircraft manufacturer as well as the supplier side. The aircraft manufacturer is the aircraft architect and integrator and holds the overall responsibility for the technical integrity and its compliance with aviation regulations. Aircraft manufacturers have therefore established stringent rules and standards to which their suppliers must adhere to not endanger the aircraft head of version certification.

During the Design Phase the agreed customization is translated into design modifications, which will trigger the detailed design of the customer aircraft version. During this phase, drawings or digital mock-ups (DMU) are generated. Design reviews are carried out in different stages to ensure compliance of the equipment (SFE and BFE) with the requirements. These reviews are performed by the aircraft manufacturer, in the presence of supplier and airline specialists. For BFE in particular, the airlines want to make sure that the specific requirements are being properly considered by the suppliers. Airlines often contract top-end design firms to develop bespoke cabin elements, which are supposed to ensure superior branding to step up competition. Major challenges can arise to ensure compliance with mandatory aviation authorities' regulations, and safety requirements when unique design solutions are implemented.

At the end of the Design Phase, all relevant drawings and DMUs are provided to enable aircraft component production start and to prepare for the aircraft final assembly, the Final Assembly Line (FAL). Suppliers' design input are required in time to allow customized engineering at the aircraft manufacturer and to produce the relevant interface and installation drawings for cabin furnishing.

Aircraft assembly is carried out step-by-step, starting with Major Component Assemblies (MCA) and converging in the FAL. The Build Phase starts before the Design Phase ends, already triggering certain supplier equipment deliveries. These deliveries are mainly standard equipment, which are not subject to major customization such as wings and high lift systems. It is common for the majority of supplier equipment deliveries linked to aircraft customization to happen in the FAL. While the aircraft sections are being joined-up, all cabin equipment needs to arrive in-time and on-quality to ensure a smooth and efficient cabin installation. This final phase of aircraft production process is very tightly managed as the contractual delivery to the airline is approaching and planning flexibility is significantly decreasing. Late deliveries or quality problems have an immediate impact on the planning and create knock-on effects on the scheduled aircraft delivery. Delivery delays can incur significant penalties to the aircraft manufacturer.

The aircraft manufacturer holds the overall aircraft design and production responsibility to ensure a consistent and efficient way-of-working between the various actors throughout the entire process, especially with suppliers involved in aircraft customization. The Airworthiness Authorities oversee the approval and regulation of civil aviation providing the Design Organization Approval (DOA) and the Production Organization Approval (POA). These final airworthiness approvals are covering all items installed on the aircraft.

4 Innovation and Customization

Product innovation in the aircraft industry follows different cycles and dynamics. Major innovations with step changes in technology like the introduction of new materials (such as carbon fiber for fuselage and wings) or new engine technologies (such as geared turbo-fans) happen at a far lower frequency than innovation in the passenger cabin, and are very often linked to the development of an entirely new aircraft model or a major derivative (for example the A320neo). This innovation speed depends on the required development and certification efforts, which can often only amortize over a very large number of aircraft. Sub-system innovations such as improved navigation systems or enhanced safety functions are considered during the aircraft program life cycle, based on market requirements and sound business cases supporting the change. Innovations concerning customer interfaces, the aircraft cabin and its amenities follow a very different cycle. Passengers are consumers and innovation in this segment must follow a much higher speed, as passengers expect the airlines to offer a cabin product, which meets current consumer standards. While an aircraft and its systems have a rather long life cycle, the aircraft cabin is subject to recurrent change. Not only passenger expectations drive this speed, but also airline strategies, which try to have its product evolve with the market, on the one hand to be attractive and competitive, on the other hand to be efficient, meaning increasing yield.

Aircraft manufacturers and OEMs have well understood this trend and its dynamic, focusing on developing a strong and competitive product offering. The particularity around the aircraft cabin lies in its multi-source business model. Aircraft systems on the other hand are often of single source due to significant development and certification efforts.

Aircraft manufacturers have a rather good control and leverage on suppliers around the airframe and its systems, which are often from single source origin. The cabin is more complex and based on a multi-source model with several suppliers for the different cabin commodities, like galleys, lavatories, seats, to mention the most important ones. This generates automatically a strong competitive dynamic, as each cabin OEM is promoting its latest products. Airlines also ask the suppliers to provide latest technological product innovations, often becoming bespoke products forming part of the airline cabin product branding. In the last years, a full lie-flat seat has become standard business class for all major carriers, first class has been pushed further to become suites, like the Emirates Private Suite (The Emirates Group 2016), or even apartment type arrangements, like the Etihad Residence (Etihad Airways 2016), which is creating a dedicated living space with living room, bathroom and bedroom. The cabin has become the battlefield for airline cabin product differentiation, and due to design agencies' creativity the airlines are pushing the limits of customization.

Challenging technologies and designs without sufficient maturity and readiness for serial production can create major supply chain hick-ups and can even end up disrupting the final aircraft assembly, especially when suppliers underestimate the required leadtime. Dynamic testing of passenger seats, for example, has been identified as one of the most frequent issues linked to certification. Another reason for project schedule slippage are resulting from design changes outside the project framework when customer requests are taken into account for product customization. As BFE supplies are developed to the specific customer requirements only, the level of standardization is low, the customer leverage is high and the risk caused by late changes is substantial. Customized bespoke products are produced in smaller series, so the effort invested in proper industrialization is limited, influencing the maturity of the finished product. The aircraft seating market is very much exposed to customization and specific designs, and is working with a supply chain of sub-tiers for electronic components like seat controls or Inflight Entertainment (IFE) systems, and mechanical components like seat actuators or seat belts. Sub-tier management is key to ensure that design evolutions are properly flowndown to sub-tier suppliers, securing the changes required to interfaces and functionalities in a timely manner. Modifications to single components can generate collateral effects on other components, putting in danger the integrity of the entire product. Changes to the screen of a seat-mounted in-flight entertainment system, for example, can jeopardize the certification of the seat and require a complete recertification of the seat.

5 Aircraft Customization: Different Approaches

Having understood the basic customization principles and the related challenges induced by airline requirements, let us have a closer look at two discrete aircraft programs and how customization has been addressed there. Airbus A380 and Airbus A350XWB are both flagship products. Certainly the A350XWB is a more recent aircraft program and lessons learnt from A380 customization have been taken into account. Airlines invested a significant amount of effort and money in its customization, especially in the cabin being the most visible part of customization, which will also be the focus of the following discussions.

The Airbus A380 is the biggest commercial aircraft in the market today, with an average seating capacity of around 555 passengers in three classes on two decks. The A380 was developed in the late 1990s, at a time, when Airbus Group did not exist yet, and the project was managed in the four European national aerospace companies following a clearly defined national work-share. In addition to the aircraft being divided into work-shares, the national aerospace companies also developed the respective work-share using their engineering tools and methods. The integration of the fuselage, which is mainly shared between the French (nose and center fuselage) and the German (forward and aft fuselage) design responsibility, was exposed to additional challenges because of the use of different design tools and methods. An integrated single data model for the aircraft did not exist, which is quite common in today's development environment thanks to the significant advances in IT tools and electronic data management.

The A380 customization was one of the very big challenges of this aircraft program. The customization of an airline flagship entered a new dimension offering full length double decks with a large staircase in the forward fuselage and a circular staircase in the aft. Premium airlines even offer private suites in the first class, or apartment type arrangements with dedicated facilities such as showers, bar/lounge areas, and trolley lifts between the decks, making the aircraft customization a project of several tens of millions of dollars. When the A380 aircraft was designed, no specific standardized provisions had been taken to prepare for cabin customization and therefore every head of version was designed and developed starting from the same starting point, being the basic aircraft fuselage. Customization is influencing the design and the development of most fuselage interfaces for cabin and cabin system installations. The design effort for structural provisions on the A380 can reach up to 70 % on average, while the remaining effort is required for visible cabin parts. Reusing design solutions is limited to the ability to build on former head of versions. As a result of the currently small A380 fleet size, the reuse level is rather limited. Relatively simple changes can easily trigger a costly redesign of aircraft structural interfaces, and might even require adaptations of wire harness routings or of other systems.

In summary, the A380 aircraft customization possibilities provide airlines an almost unlimited freedom for realizing unique cabin interiors thanks to volume and floor space, but to the detriment of highest complexity from design to final installation. Customization cost tend to be very high as the opportunities to re-use design solutions are limited because of the rather unique cabin arrangements. In comparison to other aircraft programs like the A330 or the A320 family, the customization efforts on A380 are roughly 2.5 times of an A330 or even 8 times of an A320 aircraft.

The Airbus A350XWB had been developed with the clear objective in mind to avoid repeating the challenges encountered on A380 aircraft customization, while at the same time designing an aircraft, which will enable Airbus to rapidly ramp-up production rates. During the design phase of the aircraft, it was considered to ramp-up the production ability to at least ten aircraft per month in the first few years after aircraft entry into service. Based on past experience from Airbus long range aircraft, a comparable production ramp-up took more than 15 years, while the ramp-up target for A350XWB was less than 4 years for a comparable production rate, which represents an unprecedented ramp-up for Airbus and its supply chain. To secure such an ambitious production ramp-up, one of the critical success factors to steep production increase is design stability combined with a robust supply chain delivering on-time and on quality. Complexity driven by varying product configurations and disturbances driven by (late) changes to the product design must be avoided, or at least limited, to a minimum.

The objective of aircraft customization—creating a customer specific and unique product configuration—seems not necessarily compatible with design stability and limited product changes. Customization looks at making the aircraft cabin unique in line with airline specific operating and branding requirements, allowing airlines to create a unique cabin experience and to differentiate themselves from competitors.

The challenge for developing the A350XWB, and its secret for success, was to find a smart compromise between highly standardized favoring industrialization, and sufficiently flexible solutions to enable airlines to achieve their level of product differentiation.

The A350XWB aircraft and its cabin were designed around an innovative cabin architecture. To limit perturbations of the fuselage structure design and the built process by introducing changes driven by cabin customization activities, the fuselage was designed integrating an enabling platform. This enabling platform—in simple terms—is the collection of all interfaces required between the aircraft fuselage and the cabin interior. The number of required interfaces had been intelligently defined already in the early aircraft Design Phase by working closely with the future customer community to understand their requirements concerning customization. In dedicated customer focus groups, Airbus had engaged with the airlines to jointly define the customization requirement for the various cabin commodities. At the end of this process, and after in-depth analysis and arbitration resulting also in having to discard certain requests due to their low demand or high complexity, a common interface structure had been defined, the enabling platform, which shall guarantee the necessary stability for the industrial process of building the aircraft fuselage structure. In principle, the aircraft build process should be very stable and robust, as all aircraft are being built according to the same standard, and the customer specific cabin installations start rather late in the process, at major component assembly level just prior to entering the final aircraft assembly, or even only during final assembly of the aircraft.

After the enabling platform integration, the next critical step in customization is the installation and integration of the cabin monuments. During aircraft cabin installation, the cabin elements need to be integrated from a geometrical, and from a systems point of view. Systems integration covers, for example, water supply and waste, air conditioning, lighting, power supply, but also data networks for cabin systems covering cabin operations requirements (e.g. for flight attendants to manage the cabin operations) as well as passenger requirements (e.g. wireless networks, inflight entertainment). The A350XWB aircraft cabin was addressed differently compared to former Airbus aircraft programs, by structuring the cabin in so-called door zones. Usually, door zones are areas for cabin monument installations such as lavatories, galleys, closets or even bars. In principle, there are almost unlimited monument combinations possible in these door zones. To address the design stability and ensure low disturbance resulting from design changes, predefined door zone packages were defined in line with airline requirements. These packages are again a result of arbitration of required customization flexibility and product standardization. To limit later change requests when defining customer specific head of versions, the objective was to provide a sufficient number of existing packages, which could largely cover customer demands, triggering only a few and minor adaptations. The pre-defined door zone packages are included into the A350XWB aircraft description document, offering selectable options once the pre-development has reached sufficient maturity. This way the aircraft integration is secured and the late issue discovery is limited. To further support the customer

Fig. 3 Examples of A350XWB cabin rendering based on aircraft DMU (Airbus S.A.S. 2015)

guidance process, the aircraft description document including the selectable options is incorporated into a cabin configuration tool, which allows to visualize the airline's customized cabin in 2D and 3D. As all product data on A350XWB is managed in a single DMU, the visualization is reaching a high level of detail, getting close to reality and enabling a solid verification of the considered layout and customization options (Figs. 3 and 4).

Comparing the two aircraft programs with regards to their customization approach and their supply chain implications, we can identify some fundamental differences, which are actually driven by the initial aircraft design philosophy.

- Platform Concept
- Customization Options
- Sourcing Models
- Design Tools

These main customization differences also influence the supply chain ability to efficiently support the aircraft build process. Design maturity and stability will allow suppliers to properly plan demand and industrialize production. Late changes will disturb this stable process and cause additional effort, up to the risk of supply chain failure because of missing parts. In order to allow for a stable industrial flow, the critical decision is to best define the integration of standard and customized parts. From a build process point of view, it is further important to defer the point upon which the aircraft or its sub-assemblies become customer specific to the latest possible point. By doing so, the majority of the aircraft build process is customer independent and the amount of customized parts can be reduced. The only detriment of late customization integration is linked to failure in design or in

Fig. 4 Examples of A350XWB cabin rendering based on aircraft DMU (Airbus S.A.S. 2015)

manufacturing of the customized parts, which will create disruptions at a late stage, leaving very limited time until aircraft handover to the customer. To mitigate this risk, one pragmatic option is simply to reduce the number of variations / options, and to apply modular concepts to customization, enabling re-use while at the same time reducing cost and lead-time. This strategy was applied to the A350XWB aircraft customization, by having implemented the standardized enabling platform, allowing to build the aircraft up to a certain section level independently from specific customization requirements. By limiting the customization choices, the customer specific modifications can be better controlled. Defining the sourcing strategy around fewer suppliers should allow Airbus to better control the supply chain, especially in the early program phases during production ramp-up. Large cabin parts were even single source procured as SFE. Suppliers working on the A350XWB are required to use the shared DMU and related processes, including a harmonized set of design tools. Limiting the supply base, working with the same methods and tools, and anticipating customization during aircraft design allows Airbus to reduce the customization risk.

The A380 program was not developed with these principles in mind and therefore industrialization of customization has remained very costly and complex to manage. The ambition of airlines to make the A380 aircraft unique through complex BFE projects for mainly the premium class seating has not helped to reduce or at least limit the complexity and integration challenges. A paradigm change to follow the A350XWB approach is not realistic due to the excessive cost of the change.

6 Outlook/Future Collaboration in Customization and Digitalization

Taking the lessons learnt from current major aircraft programs, we can identify several axis for future improvements. As the A350XWB program has already proven, the usage of a single and shared DMU by all parties involved in the aircraft design and build process contributes to reducing the customization integration risk. Sharing the DMU with suppliers being involved in customization enables early verification and validation of product design, helping to avoid late surprises during production or even final installation. The delays caused are very difficult to recover, if at all, and generate significant disruption and dissatisfaction for the aircraft manufacturer and the airlines.

With the evolution of digitalization of products and processes, the next step could be to enlarge the scope from pure design verification and validation based on shared DMU to functional validation and even testing of the integrated system.

Besides the aspect of leveraging digital transformation and means to improve the design process, the aspects of industrialization require more attention. At the early stage of designing an aircraft and its interior, industrialization and integration need to be considered in the sense of being designed for industrial integration. The A350XWB program has demonstrated that by implementing an enabling platform as part of the airframe, which includes already all standard interfaces, the serial production of the green aircraft up to section level can be well industrialized, as no disturbance is introduced by customization. However, when it comes to the integration of the different cabin elements they are often coming from different suppliers and are installed side-by-side into the aircraft, each element requiring its proper interfaces and supplies. Further optimization might be achieved by advanced integration of the various cabin monuments and by creating integrated architectures. This can result in additional weight reduction and improved space usage, giving the airlines more revenue opportunity, which in the end is influencing the evaluation of a given aircraft model.

References

Airbus (2015) Manage fleets – Airbus manual M20196 Issue B, Released 2 Feb 2015
Airbus (2016) Global market forecast—flying by numbers 2015–2034. Available via http://www.airbus.com. Accessed 31 Dec 2015
Aviation Week (2015) A law that changed the airline industry beyond recognition (1978). Available via http://aviationweek.com/blog/law-changed-airline-industry-beyond-recognition-1978. Accessed 13 Jan 2016
Boeing (2016) Current market outlook 2015–2034. Available via http://www.boeing.com. Accessed 15 Jan 2016
Etihad Airways (2016) The residence—on board our A380. Available via http://www.etihad.com. Accessed 11 Jan 2016

Garcia M (2014) The cost of great airline design can be measured in delays and dollars. 30 Dec 2014. SKIFT. Available via http://skift.com/2014/12/30/the-cost-of-airlines-custom-interiors-can-be-measured-in-delays-and-dollars/. Accessed 13 Jan 2016

The Emirates Group (2016) the emirates experience > cabin features> first class: private suite. Available via http://www.emirates.com. Accessed 11 Jan 2016

U.S. Department of State (2016) Open skies agreements. Available via http://www.state.gov/e/eb/tra/ata/. Accessed 15 Jan 2016

Gabriel Oehme, born in 1968, is an engineer. He graduated from FAU Erlangen-Nürnberg University and holds also a master degree in business administration from RWTH Aachen University. He has close to 20 years of experience in the aviation industry in various functions. After several years as Head of Key Account Management for Airbus Upgrade Services managing aircraft retrofit projects for Airbus' customers worldwide, he moved into Airbus Commercial being in charge of sales contracts negotiation for major leasing companies. He joined Airbus Procurement in 2008 in the indirect procurement field, where he took over the responsibility for Information and Communication Technology procurement at EADS group level (today Airbus Group) shortly after. In 2014, he joined the A350XWB Program as Head of Customer & Business Development, responsible for developing the product offering as well as the customization of the A350XWB aircraft. Since May 2015, he has been heading Airbus Programs Procurement, managing the relationship between the Airbus aircraft programs and procurement, from commercial to supply chain.

End-to-End Demand Management for the Aerospace Industry

Avinash Goré and Alexander Nathaus

Abstract Reducing the complexity in supply chain management is the central challenge for the aerospace industry and key to increase production rates and simultaneously improve the margins. A central influencing factor is the demand management along the supply chain. There is big potential in challenging the current quality and availability of demand information towards the suppliers, to improve long term forecasts and to enable efficient investments in infrastructure. Furthermore, medium and short term forecast stability is important to ensure economical utilization of production resources and to prevent missing parts respectively.

This paper focuses on the challenge of implementing an integrated End-to-end Demand Management approach in the aerospace supply chain. In the first part, we provide an overview of these challenges and propose solutions for improving long term demand based on previously known customer configurations. The second part is focusing on the practical implementation of the proposed solutions and industry examples. Based on a case study on disaggregating demand for larger suppliers, we show why the right ordering and logistics solution selection is important. Finally, we provide a short overview of how the aerospace industry can learn and benefit from best practices in the automotive industry.

We propose three solutions to enable true End-to-end Demand Management. First, the implementation of a hybrid demand forecasting. Second, improving transparency within the supply chain. Third, the selection and adherence to optimal ordering strategies.

A. Goré • A. Nathaus (✉)
Porsche Consulting, Bietigheim-Bissingen, Germany
e-mail: alexander.nathaus@porsche.de

1 Introduction

Since 2005, Porsche Consulting has been providing tried and tested automotive best practices to the aerospace industry. The mission is not to copy-paste the automotive solutions, but to use the same underlying lean principles of *Flow, Takt, Pull* and *Zero-Defects* and to adapt them to the needs of the aerospace industry.

An end-to-end collaboration is the only effective way of reducing supply chain costs for all actors in the supply chain. In a global study conducted by Porsche Consulting with >50 aerospace companies in 2014, a strategic approach for the aerospace supply chain called "The new Chain" (Porsche Consulting 2014) was developed. The study proved that companies which respect three core elements of "The new Chain" are significantly ahead of companies that neglect them with regards to productivity, cash flow and delivery performance. These core elements are greater transparency, higher quality planning, and greater industry professionalism by all partners involved in the chain.

This article will focus on practical examples on improving the demand quality, stability and availability by addressing key aspects of configuration management and demand creation, transparency creation and harmonized order strategies.

2 Current Challenges in the Aerospace Supply Chain and Possible Solutions

The aerospace supply chain is facing several challenges. In recent years, the airline OEMs have had to deal with simultaneous program ramp-ups and increased multi-sourcing. This has put stress on the supply chain and previously neglected issues in configuration and demand management have now come to light. With the production rates continuing to increase, the cost effective end-to-end optimization of the supply chain can only be achieved by a collaborative dialogue between the OEM and the supplier. Additionally, standardization, harmonization of logistics and ordering solutions as well as improving organizational capability are keys in balancing an ever complex supply chain.

In the following, we provide three solutions to tackle those challenges in order to ensure a successful demand management and improve the overall performance of the supply chain: (1) Hybrid forecasting implementation, (2) Supply chain transparency creation, (3) Ordering strategy selection.

2.1 Hybrid Forecasting Implementation: Increasing Demand Availability by Using Information on Previous Aircraft Configuration

The aircraft OEMs are known for their full order books and the health of a program which is measured by the length of its backlog. Even though the approximate sequence of aircrafts is known in advance, this information is not available in the demand propagated to the suppliers as seen in Fig. 1. While the customer forecast is provided on a medium or long term basis, specific information on the version of the aircraft remains poor up until approximately 12 months before the industrial delivery date. The main reasons given by the OEM are the insecurity in the order book based on the financial health of the customers (Dempsey 2013) and the ever changing configurations of the aircraft due to security related modifications or new options released to satisfy the pressure of decreasing specific fuel consumption per passenger (FAST 2015).

The effect of this policy means that the available forecast for a supplier is limited to the announced build rate in the long term (2 to 4 years) and medium term (6 months to 2 years). The specific aircraft version is only available to the supplier in the short term (6 months to industrial delivery date) after the customer definition freeze (CDF).

The configuration management distinguishes between standard parts which are present in all configurations and configured parts which are driven by customer defined configuration. Here we distinguish between standard options (such as lavatory at the front or back of the aircraft) and customer specific options (such as aircraft center tanks). The customer specific options usually require additional design integration and engineering effort and are thus subject to lead times worth several months.

The effect of publishing the forecast only after CDF means a very short time window available for the lead time of critical configured parts as can be seen in Fig. 2. While the forecast for standard parts is provided earlier and on a level close to 100 % of the final demand, the forecast availability for options is too short notice (as it arrives late) and also lacks reliability (as it remains on a low percentage compared to the final demand). This lack of forecast in the long and midterm leads to uncertainty in production resource planning for the supplier and usually manifests in supply bottlenecks in case of underestimation or financial difficulties in case of overestimation.

There are two classic ways of forecasting: *deterministic*, based on previous consumption and *probabilistic*, based on future demand. There are also several different hybrids that combine the advantages of both of these forecasting methodologies (Christou 2012).

In our case, we compare a purely deterministic approach (as often encountered at our clients) with a proposed hybrid of probabilistic and deterministic approaches. The deterministic approach gives the average consumption per part number evenly distributed over the entire period. The hybrid solution uses the customer slot

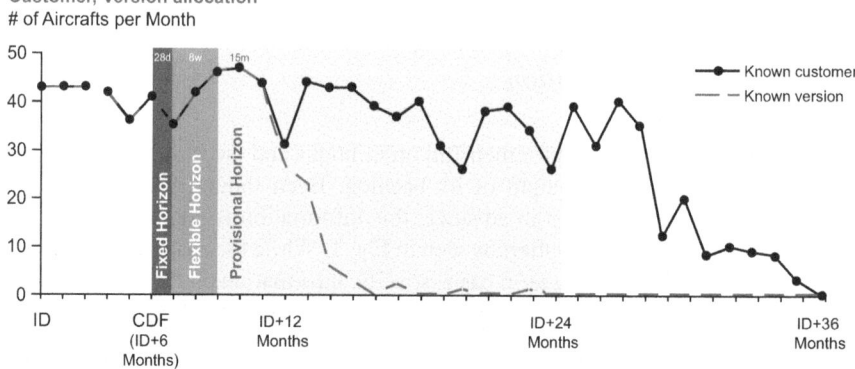

Fig. 1 The aircraft version is not allocated even though the customer is known (Porsche Consulting)

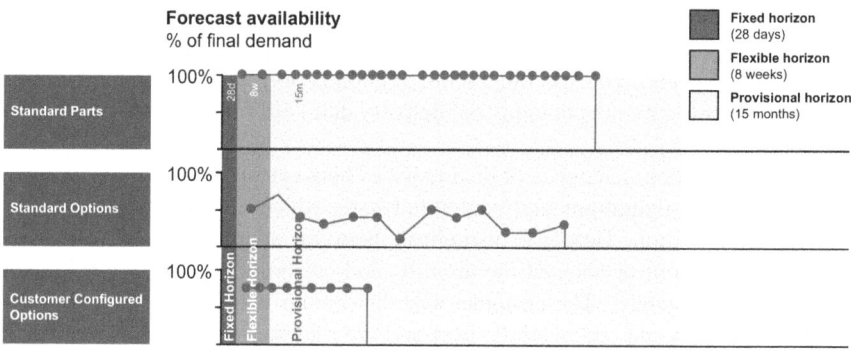

Fig. 2 Lacking forecast availability for customer options. Source: Porsche Consulting

assignment information from sales (probabilistic) in combination with the configuration information for the last known configuration used by the customer for a particular aircraft type (deterministic). This makes particular sense since the airlines have a clear incentive in keeping homogenous fleets to ease operation and service activities.

To prove the effectiveness of our approach, we compare the frequency of predicting 36 part numbers across different ATA chapters over 100 aircraft—Machine Serial Numbers (MSN). 100 % accuracy means exact prediction while a 0 % signifies complete lack of forecast and 200 % accuracy tells that the forecast was double the actual requirement. The results in Fig. 3 clearly show that the hybrid approach gives the best forecast.

Around 30 % of the aircraft sales volume is claimed by the leasing companies. Here the prediction of the most likely version is not possible since the final airline customer is usually decided quite close to CDF. Hence for leasing customers the

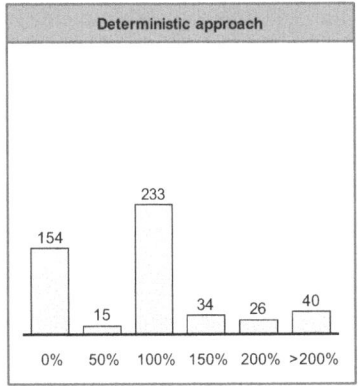 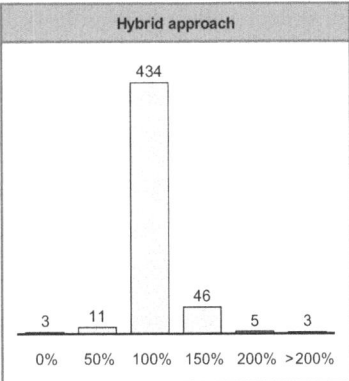

Fig. 3 Histogram of prediction accuracy per approach. Source: Porsche Consulting

probabilistic model is used, while for airline customers we recommend the hybrid model.

2.2 Supply Chain Transparency Creation: Improving Demand Availability by Enhancing Communication

As one of the major factors to achieve an end-to-end demand management, the transparency and communication among the members of the supply chain needs to be improved in order to improve the availability of accurate demand information. Therefore, the processes, dependencies and capacity availability need to be transparent along the chain. Additionally, it is also key to focus on measures to improve accurate information sharing among the different companies. To further enhance communication, it is important to share information with more subordinate levels of the supply chain by establishing standardized communication paths and increasing the frequency of data transfers (Porsche Consulting).

Standardized Communication Paths
Porsche Consulting studies among >50 aerospace companies show, that 67 % of Supply Chain Managers see the variety of different communication media as an efficiency problem. Considerable additional time and resources are required to handle the variety and complexity of media including fax, phone, e-mail, supplier portals and ERP systems. Also, every interface is a challenge for the data accuracy, in particular when manual operations are required.

Communication deficits are mainly obvious in three areas, where the negative impact of companies without uniform data exchange is approximately one third higher. First, in 71 % of the cases the companies do not know the exact extend of the actual customer demand. Second, many companies have another manual and not systemized ordering system running in parallel. Third, it is common to interject short-term orders with a high priority in the schedule on short-notice (Porsche Consulting 2014).

Frequency of Data Transfer

A high frequency of data transfer is another important factor to successful demand management. Compared to the automotive industry, the current communication frequency in the aviation industry is below average. Long term demand forecasts are just received on a monthly basis by 23 % of the companies, whereas nearly half of the companies receive them only quarterly or even less often. Even the short term demand forecasts are only received on a weekly basis by one third of the companies. When compared to the on-time delivery rates, the companies that receive demand data in high frequencies showed a significantly better performance than the companies with lower frequencies. Companies receiving demand information on a daily basis achieve an on-time delivery of 95 % on average whereas weekly information leads to only 88.75 % of on-time deliveries (Porsche Consulting 2014).

2.3 Ordering Strategy Selection: Setting Standards by Harmonizing Logistics and Ordering Solutions

While a more accurate demand forecast and better communications are a prerequisite for good collaboration, the correct choice of ordering strategies and logistics solutions are essential for reducing demand fluctuation, optimizing inventory and reducing risk of missing parts. When standardizing logistics and ordering solutions, it is essential to find a balance between keeping the complexity low while still being able to represent the required attributes of the part demand profile. It is essential to follow the *Plan For Every Part* (PFEP) strategy and a suitable segmentation method like ABC-XYZ (XYZ represent the clusters for frequency of consumption and NOT ordering frequency).

Every procurement and planning organization faces the challenge of balancing transaction costs of managing and launching POs (Purchase Orders), as well as the impact of high inventories in the ordering process of materials. The target is to achieve a global optimum for the ordering strategies, where the effort remains manageable and inventories can still be controlled to an appropriate level as shown in Fig. 4. For different industries and companies, the steepness and intersection of the cost curves vary. Automotive manufacturers, for instance, leverage a high degree of systems integration with their suppliers and automation of ordering processes to allow for an almost flat curve on transaction cost per PO, regardless of the overall material volumes. This allows automotive manufacturers to aim for high frequency ordering, demand actualization and deliveries.

The selection of the optimal ordering strategy for every part therefore depends on the balance of transaction costs and the costs of carrying inventories. Both of these costs can of course be reduced (for example IT platform integration of suppliers, digital PO workflows or reducing warehouse costs).

End-to-End Demand Management for the Aerospace Industry 111

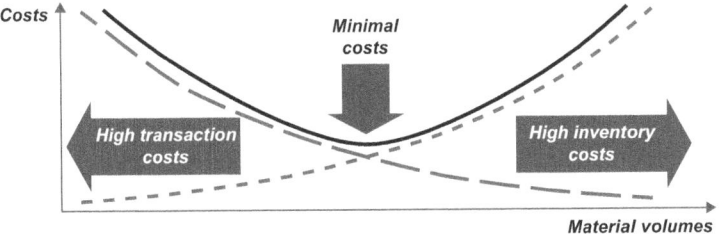

Fig. 4 Balancing of transaction costs vs. inventory costs. Source: Porsche Consulting

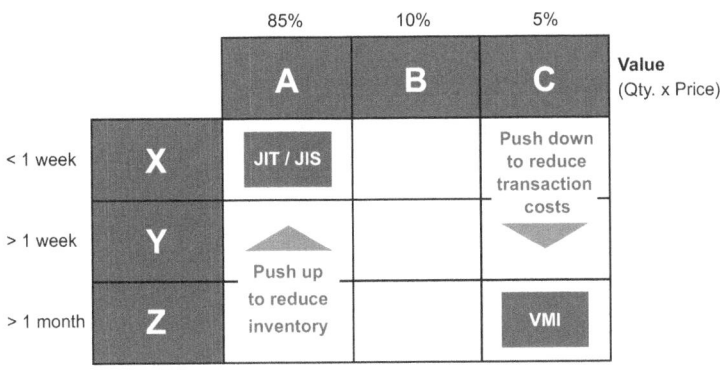

Fig. 5 Schematic figure of a typical ABC-XYZ matrix (Porsche Consulting)

To avoid micromanaging, every material number and material categories have to be established. Each material category is then allocated with the optimal ordering strategy. The general logic is shown in Fig. 5. Even though these principles always hold true in any industry, there are particularities to be considered to achieve the best supply chain fit with the ordering solutions. An automotive manufacturer, with its comparatively low transaction costs, will have a much higher emphasize on ordering as many materials as possible JIT (Just in Time) or JIS (Just in Sequence) to minimize inventories. Aerospace clients, with their comparatively high transaction costs, often focus on reducing their ordering effort and therefore prefer vendor managed inventories and achieve only a small share of true JIT or JIS deliveries. Due to the negative impact on inventories and eventually shipset costs (Vendor Managed Inventory is not free), we recommend to invest in processes and infrastructure that allows for low transaction costs and therefore more inventory focused ordering solutions. Not necessarily JIT or JIS (as these are complex to achieve), but rather well managed KANBAN inventory control system or lot size ordering points with a low as possible inventory coverage.

3 Practical Examples from Aerospace and Automotive Industries

There is an opportunity to apply best practices for demand management from the automotive industry within aerospace to improve the overall supply chain performance. We want to highlight a few examples how this can be achieved in the following case study. The case study is based on real supply chain projects within the aerospace community.

3.1 Case Study: Disaggregation of Demand for a Large Supplier

The supplier in question was delivering a large amount of part numbers with mainly standard parts or standard options. The main complaint by the supplier was that the demand fluctuation seen by him was much higher than the actual fluctuation in the aircraft build rate. This was validated by comparing the build rate to the parts needed within SAP and the purchase requisitions and orders sent to the supplier as shown in Fig. 6. Taking the calendar weeks 24 and 25 as an example, the purchase orders dropped by -54% while the aircraft build rate (-17%) as well as the needed pieces (-15%) had a much lower fluctuation.

Further investigation showed that the ratio of optional parts was not high enough to explain this demand fluctuation. A look at the MRP settings revealed the real culprit. The lot size was set to represent the monthly consumption. This meant that most of the parts were ordered in the first week of the month and this resonance was creating demand peaks which meant the supplier was required to deliver up to 40 % of the parts in the first week of the month.

The implemented solutions reduced the demand volatility from a supplier perspective notably. The delivery frequency was already twice a week, so there was no reason to reduce the lot size further. In addition, planning calendars (SAP 2015) were set up to disaggregate demand and achieve fixed interval ordering resulting in a pattern which was easy to follow for the supply officer.

This resulted in decreased demand fluctuations and reduced average weekly stock as seen in Fig. 7. The bundling of transport volumes as seen in Fig. 8 to reduce transport costs is an additional effect of planning calendars and can be used in transport network optimization and goods receipt levelling.

It is important to note that the reduction in lot size is often not possible due to contractual obligations. Here planning calendars are the most effective way of reducing demand fluctuations by disaggregating demand.

Figure 9 shows the final results achieved by the change. It is important to note that the number of purchase orders—used as an indicator of supply office effort in some industries—stayed almost the same.

End-to-End Demand Management for the Aerospace Industry   113

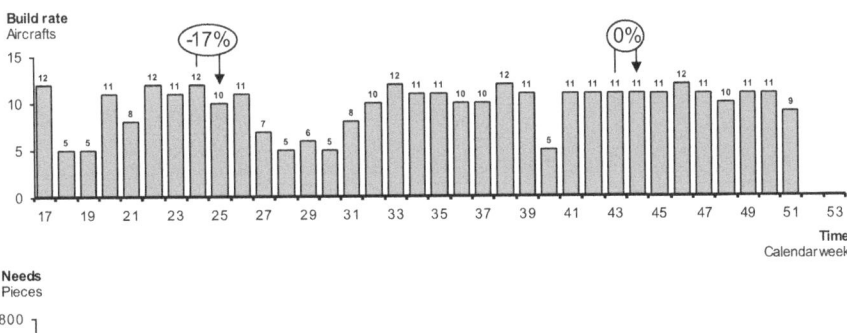
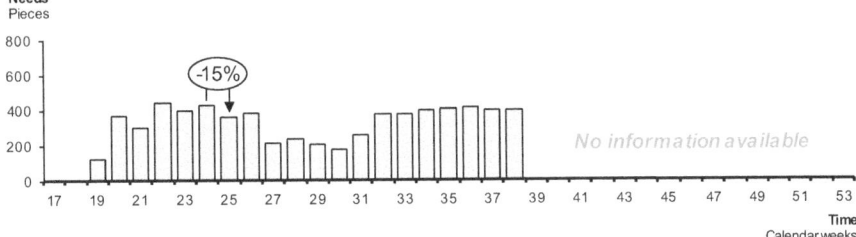
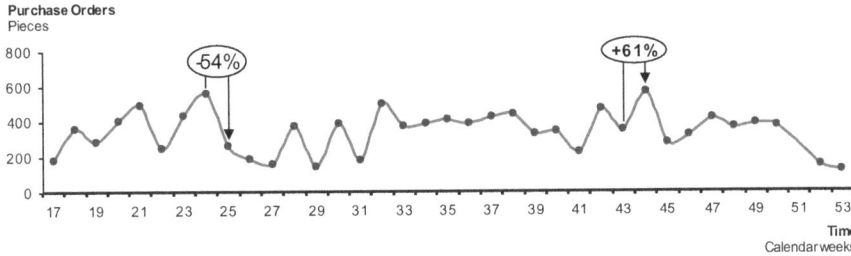

Fig. 6 Demand fluctuation in purchase orders (Porsche Consulting)

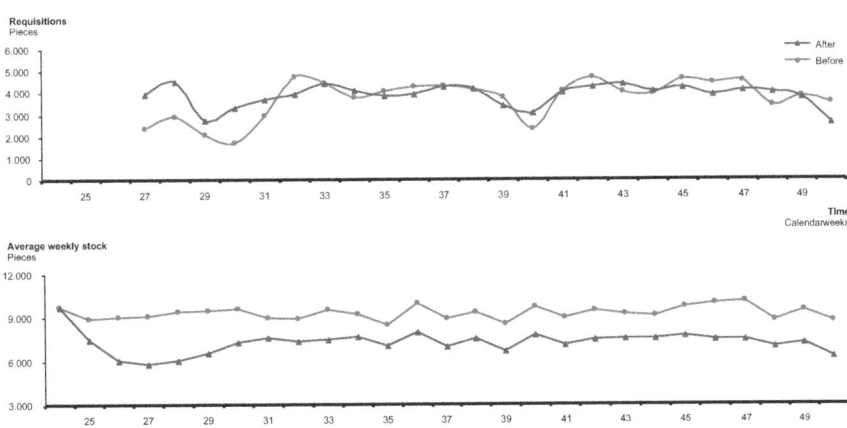

Fig. 7 Reduced fluctuations and average weekly stock (Porsche Consulting)

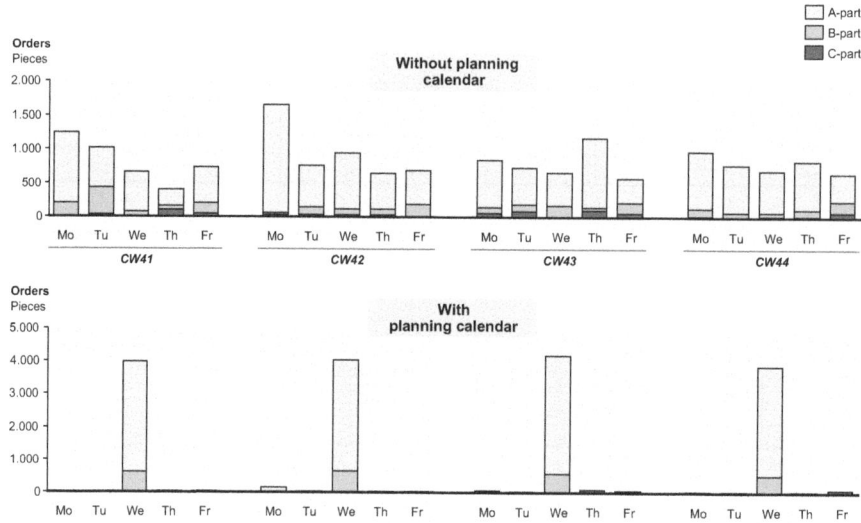

Fig. 8 Demand bundling effect of planning calendars (Porsche Consulting)

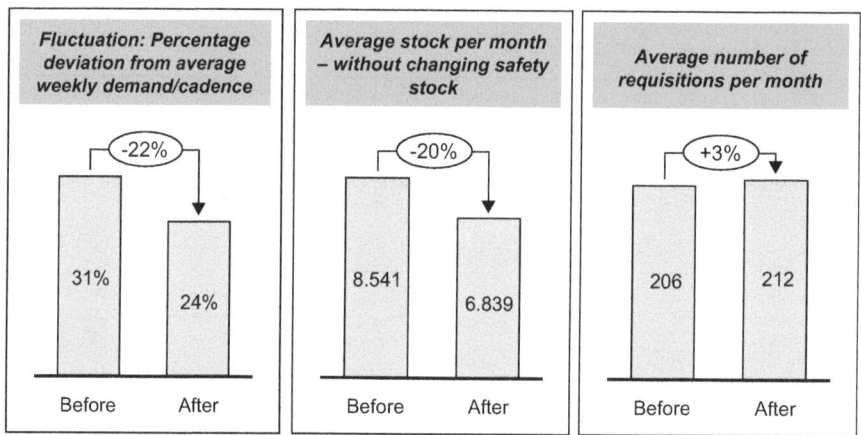

Fig. 9 Improvements without increasing effort (Porsche Consulting)

The main learning from this case study is that the choice of the right ordering solution needs detailed analysis at part and supplier level, active collaboration with the supplier and proper training for the procurement organization.

4 Looking Beyond: Benchmark Automotive and Aerospace

We see two yet untapped potentials to achieve further improvement in End-To-End Demand Management within the aerospace industry, which are also fairly recent developments within automotive.

The first potential is the collaboration of procurement and the transport organization (also transport service providers) to achieve an optimization of delivery modes across the transport networks. This is a natural evolution of using planning calendars and based on our project experience can reduce transport costs on individual routes by 30–50 %. This is achieved, for example, by bundling pickups and transferring these transport modes from very expensive "Integrator" (sending individual parcels) to "Bulk" deliveries. This is an optimization that neither the supplier nor the transport organization (or provider) can achieve without the collaboration of procurement.

The second potential is a coherent application of "category management" throughout the procurement organization. Instead of departmentalizing the procurement organization by suppliers or material groups, a stronger emphasize can be put on technologies and manufacturing techniques. This allows the buyers to develop strong expertise in a defined field (e.g. composites) and enables him to understand supplier limitations but also better judge their competitiveness.

Another promising development can be expected with the ongoing digitalization of industrial processes. The digitalization of the supply chain and the positive impacts of *Industry 4.0* on the demand management can be an enabler for a stronger integration of the supply chain. Therefore, several use cases can be examined. First, the generation and distribution of greater amounts of specific data and its utilization can be of great value for optimizing the end-to-end approach and to achieve faster or even real time data exchange with a greater accuracy.

Additionally, "Big Data" can be used to analyze historical demand values to a greater extend and optimize future forecasts. The new possibilities to analyze greater volumes of data can be of great value for detailing the analyses executed, even up to the extent that accurate forecasts might be available on part level.

Predictive analysis as enabler for greater supply chain stability based on Big Data could be another field for future research. Especially the prediction and thereby anticipation of critical future events such as likelihood of production delays or quality issues that might cause delays or backlog/missing parts can be of great value for guaranteeing a stable supply chain.

5 Conclusion

As a concrete recommendation, we believe the aerospace OEMs should take the lead in improving the demand management approach throughout the supply chain. The overall performance of the supply chain is key to improve productivity, margins and delivery performance.

Therefore, a first step would be to apply the hybrid forecast approach in order to improve the forecast quality throughout the supply chain. The target is to use the customer slot assignment information from sales (probabilistic) in combination with the last known configuration used by the customer for a particular aircraft type (deterministic). As the forecast needs to reach every step in the chain, the level of transparency needs to be improved by enhancing the information flow. This needs to happen on two levels—the standardization of communication paths and the increase of the data transfer frequency. As a last step, a suitable ordering strategy needs to be selected for the specific materials and parts ordered from the supplier. This is essential in order to prevent demand fluctuation as well as to optimize the inventory levels and to reduce the risk of missing parts.

We see the contribution of "Industry 4.0" within End-To-End Demand Management within real-time exchange of demand and consumption data as well as predictive analytics of possible supply chain disturbances.

6 Porsche Consulting and the Aerospace Industry

Porsche Consulting has successfully completed more than 200 projects in the aerospace industry. With 40 experienced experts in the industry, the international consulting team focuses on five service offerings in the manufacture, in the maintenance and in the operation of commercial aircraft: (1) *Reliable interplay in the supply chain,* (2) *Knowing, using and further developing core areas of expertise,* (3) *Accelerating production rates,* (4) *Reducing inventory levels,* (5) *Airborne again promptly.*

The focus on improving supply chain coordination can reliably increase the efficiency of even highly complex structures as the automotive industry clearly demonstrates. Three key factors are essential: transparency in the processes, better quality planning, and a higher level of professionalism based on a new understanding of roles.

Company analyses indicate to necessary changes to reach the set goals and to develop a strategy for product portfolio application. Aerospace business often grow by means of acquisitions and are therefore strongly marked by divisional lines between, for example, maintenance, defense or aerospace. This is associated with a tendency to think in compartmentalized business fields. The expertise available in different divisions is often not networked above and beyond these divisions. Porsche Consulting's analysis generates a valuable overall view of available technologies and areas of expertise. This enables resources to be matched to future demands.

International airlines are ordering considerably more aircraft than manufacturers worldwide can currently produce. This puts agreed-upon supply deadlines at risk. When manufacturers respond by ramping up their production, they need the highest possible degree of stability. With suitable strategies, production rates can be increased by 10 %.

Inventory levels are deceptive and give the illusion of security. While it might seem hard to reduce them at first, doing so only brings benefits. Lower inventory levels mean more capital for investments and a greater ability to respond to market developments. Measurable results can be produced by means of active inventory and range management. Experience in the aviation supply industry shows that up to 25 % of inventory levels can be reduced for raw materials and semi-finished products. For end products, reductions of up to 15 % have been demonstrated.

Aircraft only make money in the air. Ideally, planes would fly 365 days a year. Every malfunction negatively affects punctuality and therefore also the airline's profitability. With an industry-oriented approach, Porsche Consulting questions and examines established modes of maintenance processes, luggage conveyance, passenger processes, and airplane dispatch. In maintenance, for example, our sophisticated methodology reduces down times by 20–30 %.

References

Christou I (2012) Quantitative methods in supply chain management. Springer, London
Dempsey PS (2013) Airline bankruptcies: the post-deregulation epidemic. McGill University, Montreal
FAST Airbus Technical Magazine (2015) Incremental development. Available via http://www.airbus.com/support/publications/. Accessed 23 Jun 2016
Porsche Consulting (2014) The new value chain—greater efficiency in the aviation indus-try. Available via http://www.porsche-consulting.com/en/downloads/?tx_xxmultimedia_pi1%5Bfile%5D=613&tx_xxmultimedia_pi1%5Bcontroller%5D=Multimedia&tx_xxmultimedia_pi1%5Baction%5D=download&cHash=318e7754924394b13b8cdce733f1565c. Accessed 23 Jun 2016
SAP (2015) Planning Calendar Material Requirements Planning (PP-MRP). Available via SAP (2015) Planning Calendar Material Requirements Planning (PP-MRP). Accessed 23 Jun 2016

Avinash Goré, born in 1979, has an engineering degree from Chalmers University of Technology, Sweden. He has 12 years of experience in a broad range of fields from engineering, production, logistics and supply chain management. Initially working in research at Robert Bosch GmbH, he moved on to Supply Chain Management at Henkel KGaA. Since 2010, he has been working with Porsche Consulting GmbH in automotive and aerospace practices where he has led several projects in supply chain and production optimization. Since 2013, he has been managing the Airbus commercial key account.

Alexander Nathaus, born in 1985, has degrees in Finance, Investments and Environmental Technology from ESCP Europe in Paris, France. He has 6 years of experience in the automotive and aerospace industry. Before joining Porsche Consulting in 2011, he worked in the financial industry in London and Sao Paulo.

Main Differences and Commonalities Between the Aircraft and the Automotive Industry

Horst Wildemann and Florian Hojak

Abstract The aircraft as well as the automotive industry have changed tremendously in recent years. This has mainly an impact on sales market globalization, increasing product individualization and product development process standardization. Strongly differentiating production sites are forming a global network, stronger product differentiation and transferring added values to suppliers. A new trend is the additional focus on design, development expertise and production coordination. Both industries have high quality standards, efficiency, individualization and innovative technology in common, raising the question: what are the chances to learn from each other's production and supply chain processes?

Figure 1 provides an overview of the main differences between the two products and their production.

The differences between Aircraft and Automobil can also result in various similarities such as product individualization for customers causing significant alterations, the aim to reduce delivery time and to streamline production and logistics processes by using a large number of nonvariables, assembly group and site modularization, improving factory utilization and a transition to a flexible production line, supply chain control as well as production process digitization in early stages of the product development and throughout production. Both industries are challenged by similar trends.

1 Trends and Main Fields of Action for Production and Supply Chain

Over the past few years, the automotive industry has been challenged to 'reinvent' the automobile. Assignable causes of the reinvention are measures to reduce CO_2 emission, alternative drive concepts such as hybrid or electric engines and changing

---

H. Wildemann • F. Hojak (✉)
TCW Management Consulting, Munich, Germany
e-mail: florian.hojak@tcw.de

global market conditions causing companies to relocate their production sites to newly industrialized countries (NIC). Ongoing trends have to be taken into account, too. Control systems are digitized supporting the development of functions that increase vehicle and customer safety and infotainment programs are established. Light weight constructions using new materials such as carbon or magnesium are common standard. Connecting the vehicle with communication systems shall finally prevent accidents and support traffic routing. When it comes to alternative drive concepts in particular, a dominant development direction has not yet been established. Consequently, car manufacturers have to simultaneously offer a variety of product technologies causing massive research and development cost (Meißner 2013). The automotive industry operates globally, thus being able to produce complex products containing up to 18,000 components using equally complex processes. The industry produces enormous quantities while still being able to react to individual customer preferences. To enable these production processes, a number of stakeholder groups such as component, module or system suppliers and different service providers participate on different levels (0.5–3) of what's called the 'Tier-pyramid' (VDA-Arbeitskreis PLM 2012). On this basis, the following key fields of action are not only identified for the automotive industry, but can also be of importance for Airbus (Fig. 1).

1.1 Improving Energy Efficiency in Drive Concepts

A concept promoting engine efficiency enhancement consists of downsizing the engine by reducing cubic capacity, turbo and high pressure injection, cylinder deactivation and variable valve lift (VVL), 8 or 9 speed automatic transmission, combustion chamber and friction optimization as well as light weight constructions. Optimized energy management with increasing on-board voltage from currently 12–48 V is also part of the enhancement program, just as decreasing cable sizes to reduce weight and costs while increasing available space. Cable capacity has to be raised from 3 to 12 kW to gain brake and drive power through hybridization. In the aircraft industry similar optimization processes are taking place improving jet engines, materials and production methods. For the A320neo, Airbus is using a geared turbofan built by P&W to reduce energy consumption and noise at optimal rotation speed of both fans and low-pressure turbines. The new turbine generation is already in use in the latest generation of aircrafts.

1.2 Developing Alternative Electric Drives

Established car manufacturers and new competitors are currently working on hybrid and electric drive solutions. New competitors such as TESLA, Google and Apple will most probably use altered value chain concepts without being directly

		Aircraft (A320)	Automobile (Audi A3)
Product	Power (HP)	> 66.000	> 100
	Maximum speed in km/h	~ 900	~ 200
	Size in m (length * width * height)	~ 40 * 34 *12	5 * 2* 1,5
	Model portfolio (without derivative)	~ 4	~ 12
	Possible combinations for customer equipment levels	~ 10^{34}	~ 10^{36}
	Development time until SOP in years	~ 7	~ 4,5
	Customers worldwide	~ 400	~ 100.000
	Number of components (pieces)	~ 3.000.000	~ 18.000
	Selling price in EUR	~ 90.000.000	~ 27.500
	Operational life in years	~ 40	~ 9
Production	Fuel consumption per 100 km and person in liters	~ 3	~ 5
	Empty weight in tons	~ 60	~ 1,6
	Production capacity per year	~ 260 (Hamburg)	~ 130.000 (Ingolstadt)
	Shifts per day	3	3
	Assembly time in hrs. (only final assembly line without aircraft painting)	~ 720	~ 21
	Depth of added value (equity ratio) in %	~ 30	~ 27
	Number of workers per shift	~ 500	~ 500
	Production type	Mini-series production	Mass production
	Suppliers	~ 2.000	~ 110
	Logistical expenditure per product in percent of the selling price	~ 5	~ 1,5
	Staking points / weld points	~ 1.000.000	~ 5.100

Fig. 1 Product and production differences between aircraft and automotive industry. Source: Data has been gathered by Airbus and through various TCW-projects in the automotive industry (2016)

involved in production. Cooperations with firms like Magna, Valmet or The Netherlands will cover the production process, leaving key competences such as marketing, sales and service with TESLA and companies like it. Every car firm is currently focusing on enhanced batteries with higher power density. Airbus is also looking into electric concepts for aircrafts. One of these disruptive innovations might be an aircraft called 'E-Fan' which was able to cross the English Channel solely by using electric power when first tested. Research to further apply this technology in larger aircraft has been conducted within the project 'E THRUST'. The new electric power unit has to be cleaner and less noisy than previous jet engines. Great challenges concerning battery weight and power density need to be overcome in order to unleash the technology's great potential. The technology's success will not only be subject to development outcome but also depends on the airline industry's judgement.

1.3 Technology and Premium Brand Leadership Because of Large Investments

The automotive industry mainly focuses on what's called a 'Smart Factory'. The term refers to a vision of a self-organizing production environment where production plants and logistics systems will interact with minimum human interference. The technological requirements are based on cyber-physical systems, which communicate via the Internet of Things. Part of that vision for the future is a functioning communication between product and factory (Munde 2015). The product delivers relevant information for the manufacturing process in accessible form, for instance, on an RFID chip. The factory will be able to evaluate the necessary manufacturing steps based on this data and will put them in best order. Airbus is developing a similar concept with equivalent goals under the name 'Factory of the Future'.

1.4 Extending the Product Portfolio

Besides SUVs, large limousines and derivatives, the automotive sector is developing new vehicles with electric drives. Tesla's Model S and BMW's i3 are both examples for in-house-developed electronic vehicles that have already entered the market. This new generation of cars requires new suppliers, partnerships for resources, batteries, electric engines and control systems. Airbus is also successively increasing its portfolio. The 320-product line for short- and medium-haul destinations has been continuously extended and renewed. For long-haul distances the A330, the A350XWB and the A380 represent a broad product range relying on a variety of suppliers. On top of a raising number of customers with individual requests, the complexity and product range is increasing along with the supplier's portfolios.

1.5 Extending the System Strategy

Standardization and modularization in the automotive industry has helped to exploit significant saving potentials. The Volkswagen Group's Toolkit Strategies 'Modularer Querbaukasten' (MQB) and 'Modularer Längsbaukasten' (MLB) have saved the company billions of euros by making some investments no longer necessary, while reducing part diversity and the number of suppliers at the same time. Less effort is needed for coordination, which reduces product and process complexity. Airbus has shown first standardization attempts by standardizing all parts except the fuselage's length and door specifications for the product lines A319, A320 and A321. The same concept has been applied to the A350-900 and the A350-1000. A concrete example for modularization is the A350XWB which contains modular kitchens, displays and switches in cockpits as well as modular cabins, painted fins and prefabricated modular wiring that is installed in the aircraft's ceiling.

1.6 Exploiting Emerging Markets Through Allocating Development and Production Capacities

The production capacity in the automotive sector has increased significantly within the past decades. Constantly rising customer numbers in the BRIC countries plus foreign exchange and logistics risks have caused this development. Local content requirements have to be met, forcing companies to establish additional production locations. Established suppliers have also become more international for two reasons. First, products of local suppliers often lack certification and second, suppliers have to reduce their production costs due to global cost pressure. A similar procedure can be noticed at Airbus international production sites in the United States and China. Additionally, new competitors from China (COMAC), Russia (Sukhoi) as well as existing ones from Brazil (Embraer) and Canada (Bombardier) are pushing into the regional jet market encroaching on Boeing's and Airbus lowest product segment.

1.7 Digitization and Visualization of Product Development and Production

The automotive industry has used information technology in all development process phases for decades. Vehicles are visualized in the early development stages, mechanical stress is simulated, aerodynamic characteristics are measured, manufacturing and assembly is supported and clients can use configurators on the Internet. Along with a strong network supporting knowledge exchange and the solution process, the connection between human and machine has become more and more important. In the production process of the Audi A3 in Ingolstadt (Germany), robots solely manage the luggage compartment lid and door assembly. In the aircraft industry digitalization and visualization have also been common methods used for decades. The interaction between human and robot on the other hand is still in its infancy. Weld seams located on the pylons of the A330neo, the A350XWB and the A380 are carried out by robots. Riveting robots are also used for the fuselage joints of the A320 product line. Aviation experts see great potential for further applications in production and for periphery areas.

1.8 New Business Models

The automotive industry is extending its services following the premise 'using instead of owning'. That includes constant consideration of viable business models (Hojak 2015). In this context automobile manufacturers have mainly established financial and car sharing services. BMW's DriveNow car sharing service can be found in several cities all around the world, offering customers premium BMW and MINI driving experience. At Airbus, similar ideas are becoming more relevant such as the concept of maintenance rates based on flying time ('power-by-the-hour') or aircraft leasing and manufacturer-oriented jet engine maintenance including real-time operation and analysis. Changes or service extensions aim at minimizing flight operating costs (Alcock 2015) are (just as 'DriveNow') an alteration of the original business model.

These examples show that both industries face similar challenges. In terms of production it means higher complexity in several areas such as quality assurance, logistics, staff assignment and coordination, as well as supply chain focusing on just-in-time and just-in-sequence applications. The previously named key fields of action have great influence on quantities, production and logistics costs and expenses to ensure quality. The main question is: what concepts are suited and transferable to fulfill the key fields of action?

2 Solutions in the Automotive Production and Transferability

To keep or even extend a competitive advantage in the automotive industry, several state-of-the-art concepts have been established for product and production optimization, product development and assembly. These include: multi-product and multi-brand plants, production segmentation, standardization and modularization, ramp-up and change management, logistics and order picking optimization, continuous improvement process 2.0 as well as digitization and visualization.

2.1 Multi-Product and Multi-Brands Plants

Achieving high earnings and a high degree of production plant capacity utilization are challenges in the automotive sector caused by variant diversity and localization. To overcome these challenges, vehicles are produced in multi-product and multi-brand plants. The VW-concept is used in approximately 50 % of all their production plants and aims at optimizing the flexibility of both the whole factory and the production line. The goal is to produce 'bumper after bumper'. To achieve the maximum operating grade, technical characteristics are already considered during product development to ensure smooth production. The basis is the Modular Toolkit Strategy. Innovative products and technologies can be applied to all brands, thus minimising redundant repetition of the same development. This requires concept consistency for design and construction to follow uniform principles using the same joining technology and assembly sequences. At the multi-brand production location in Bratislava, vehicles from several brands (Volkswagen, Audi,

Porsche, Skoda, Seat and Bentley) are produced. The aforementioned uniform application principle successfully creates a platform for the Sports-Utility-Vehicle (SUV) line. Non-characteristic materials and joining technologies are unitized for the VW-Touareg, Audi-Q7 and Porsche Cayenne. For example in the past, Porsche's wheel apron panels were attached with screws while Audi used clips. Clips are now used for both models in the factory. Further alignments were introduced for the assembly and attachment process of doors, center consoles and both front- and back-end-parts.

With the Bratislava plant establishment, VW set up a prime example for a multi-product and multi-brand plant. One production line similarly produces multiple models including the VW Touareg, Porsche Cayenne, Audi Q7 and Bentley Bentayga. A smaller line covers production of the VW up, Seat mii and Skoda Citigo. Admittedly, however, the 300,000 vehicles produced annually on the 'small line' are very similar, considering that they belong to different brands. Although they all have the same body construction, the differences between the SUV models are brand-specific and equipment levels are clearly distinguishable. Savings in investments for research and development and production are around 0.8 billion euros annually.

At both the Airbus' plant in Hamburg, where the A320-line is located, and at the Toulouse plant, where the A330 and the A350XWB line is located, similar conditions can be found. Both plants aspire toward flexible and versatile production while still being 'lead factory' for other plants. Soon the factory in Hamburg will manufacture the whole A320 on four production lines with a potential monthly output of up to 32 aircraft. That makes the Hamburg-site a 'lead factory' for production plants in the United States and China. Airbus is planning to produce wings for the A330 and the A350XWB on a 'Mixed-Model-Line' in Bremen. Transferability is clearly given and present efforts at Airbus can be further increased.

2.2 Line Production and Production Segmentation

There is an increasing number of product variants because of individual customer requests and regional specifications occurring in line with globalization. Strategic fast tracking to increase profits lead to capacity constraints, non-competitive through-put times and quality defects, all of which cannot be compensated within present manufacturing structures (Wildemann 1997). One possible way to introduce a reliable process is establishing production segmentation (Wildemann 2016a). Technological innovations, delivery reliability and readiness for delivery are what separates market players into 'Firsts' and 'Followers'. In order to provide good service, suppliers have to build up their stock, which causes additional costs hence lowering their profit. Assuming functionally oriented production, the portfolio has been divided into market-oriented segments while capacities have been assigned accordingly. Segmentation divides 'top-sellers' and 'exotic products' with both groups having their own production lines. Pre-production costs can be lowered and planning is more reliable. The production layout design has improved transport distances and material transfer. The guiding principle for factory design focusing on logistics is aligning the advantages of both line production and shop fabrication. This allows cost benefits and productivity advantages on the one hand and flexibility on the other hand. The goal is to utilize capacities.

The concept creates 'factories in a factory' which will unleash potential to create competitive advantage by focusing resources on a specific production task. 'Factories in a factory' cover several logistics chain steps and indirectly integrate production-related functions that make these distinguishable. In the automotive industry, for instance, a car can be produced in 80–120 assembly steps depending on the car and its equipment level. To make that possible, 'Feeder Lines' were built in order to assemble certain components and transport the right quantity to the main production line at the right time. The delivery can follow the just-in-sequence principle being connected to the main production line. The door assembly is a prime example. First, the doors are disassembled and brought to special assembly lines on which each door gets reconstructed specifically for each car model. Afterwards the completed doors are refitted and integrated in the car. The process shortens the cycle-time while less labor is required. The output can be increased by 18 % in total. Airbus uses the same method for delivery, comissioning and seat installation. 'Feeder Lines' are used to prepare the seats and deliver them to assembly at the right time. This segmentation can be applied to Airbus production and could be considered for further applications.

2.3 Standardization and Modularization

Profitable production of a growing number of various vehicles calls for higher flexibility and production resources efficiency considering the whole life cycle from building and running a production site to altering or expanding it. Multi-product and multi-brand factories are needed for small-scale model production. Because of shorter innovation and product life cycles, new car model and derivatives production must be integrated into the existing shop floor more frequently. Time and effort have to be minimized if modifications or new installations are necessary to produce new models. This can be achieved by using robust and well-proven production means. During the operating stage, certain production lines are to be construed in order to ensure flexibility with regards to different models or variants. When Audi started building the plug-in-hybrid-model A3 e-tron, its production was integrated into the conventional A3 production line. Installation engineering and employees' knowledge and quality of work level need to account for ambitious quality objectives. In consideration of the circumstances, the VW Group has based its production concept on modularization to ensure quality and stay flexible and profitable at the same time.

VW's 'Modular-Toolkit-Strategy' aims at standardizing organizational elements, production resources and processes across all production locations (Waltl and Wildemann 2014). For the purpose of reaching the whole potential of standardization, the logical order of modular production structure has to follow the 'Modular-Toolkit-Strategy'. The logical order was developed using a pragmatic approach. Modules were defined based on existing investment groups, component lists, and best practice application gathered from different production sites as well as prior-existent corporate and brand standards. The production capacity, the combination of materials, the mechanization degree and the flexibility are the determining factors that set boundaries and determine the layout design for each module. The production capacity describes the maximum amount of vehicles that can be produced on one production line and is typically specified in vehicles per hour. The material combination refers to the specific use of various materials and which of them are used together. The mechanization degree characterizes the mechanized process proportion compared to the total amount of individual processes. The term flexibility refers to the amount of product variants that can be manufactured using the same module. The combination of the determining factors equals a basic module's variance. Its implementation is only realized to a degree that is in alignment with a rational modularization concept. The possibility of changing the module, however, needs to be kept in mind and its technical design needs to reflect that. The basic modularization model covers every production level including the factory and its assembly section, different manufacturing segments and facilities, the same as single work stations. Production areas such as manufacturing, assembly, logistics or information technology and different planning functions are other considerable dimensions.

The 'Modular-Toolkit-Strategy' follows a top down approach. A first level module consists of several modules from the second level. Every second level module involves a certain important factory issue such as technics, structure, information, logistics and organization. The third level is about the assembly section which each module technics refers to. That includes specific construction in subsections, for instance, the press shop, body construction, paint shop, assembly hall and pilot plant as well as systems that spread over more than one assembly section. The 'Modular-Toolkit-System's logical order mainly depends on the modules of the specific production facility. These modules of the specific production facility are independent pieces of the toolkit system and include a certain technical scope, functions, interfaces and contents. There are factory, product, and variant-specific requirements for the modular toolkit system. Due to those requirements, modules have to be exchangeable, combinable and extendable (Waltl and Wildemann 2014). Also organizational responsibilities are defined for each model. While the modular toolkit systems' conceptual design and piloting are treated as a project, responsibilities concerning coordination and administration are carried out by a line organization. It contains a coordinating office, special approval boards, module experts for several areas, officers for certain packages of central areas and officers for certain fields of action in the factory. In order to implement the modular production toolkit, clearly defined administration and coordination processes are utilized to release or amend specific modules. Beyond that, it is characterized how the modular toolkit system is used in product development and model upgrading. An order of steps is clearly defined. Conceptual design is followed by a piloting, coordination, administration and finally implementation and controlling phase. Factory, section and product specifics need to be considered during the implementation process. The whole process ensures that every requirement can be fulfilled by a production ramp-up proven module. On the one hand the modular toolkit system can help to enhance quality, knowledge and flexibility, on the other hand time and effort can be reduced. Process innovations based on solid production modules, standardized interfaces, defined functional scopes and successful pilot phases can be integrated in the module catalogue. A production module's implementation is carried out gradually per vehicle project. Problems and errors are gathered and fixed during the process. Lessons learned help to successively adapt to standards. In that way, know-how is preserved. Easy access to information, minimal professional training requirement, fast repair and reducing the error ratio are necessary to gradually implement process innovations in the global module catalogue. Modularization in connection with technological change management in production can be crucial for connecting and digitizing production processes.

The Volkswagen Group is employing the Modular Production Toolkit to embrace Industry 4.0. Resources, products and components are clearly identifiable during vehicle production, their as-is state is describable and their final state can be programed. The production modules are vertically connected to economic processes. Horizontally, they are linked with value added networks which can be controlled at all time. An automatic production system with a worldwide network can be created by developing and maintaining a modular toolkit system connecting divisions, businesses and companies. The possible savings affect investments and professional training. A company can save almost five billion euros over several years, which reflects great saving potential. Even though the A330, the A350XWB and the A380 have segments of the fuselage aft section that can be manufactured with a modular system, the difference between modules of different airplanes is significant. It seems difficult for Airbus to transfer results from the automotive production in this respect.

2.4 Ramp-Up Management

Ramp-up is of great relevance for both the aircraft and the automotive industry. In the aircraft industry penalties have to be paid if deadlines for handovers are not met. In the automotive industry, series production starts on a date that has been agreed upon 4½ years before. Marketing and sales concentrate all their resources on that event. Waiting periods for preorders over a few weeks or months are not acceptable. If customers do not receive their product on time, it is highly likely that they will cancel their order and reach out to a competitor. Customer recovery costs 55,000 euros on average. Assuming a large-scale batch production, there is not a single vehicle that could cover that amount with its profit margin. Fast ramp-up is conclusively key to both industries. In the automotive industry it is a very good sign if 80% of the new car production capacity is reached within 8 weeks (Wildemann 2016b).

Normally that process takes twice the time. In the aircraft industry it takes 12 months to reach that target value, taking into account the high degree of individualization. Lessons learned and backflow play a great role in aircraft production ramp-up. To account for a high individualization degree in the automotive industry, the pilot production center is the link between the early stage during product development and the actual start of production (SOP). The center's purpose is to prepare for batch production. To do so, it connects the pilot vehicles and the technical development of the pilot hall on an organizational and process-related level. There are several tasks the pilot production center needs to handle. It needs to plan and guide vehicle projects through the prototype and pilot phase and influence product development in terms of economic and technical feasibility with respect to mass production. It also needs to provide test objects and accounts for virtual product and process security for both prototype and series production stage. It needs to assure that test vehicles and pilot series are flexible and efficient. Finally, it is the center's responsibility to work on a smooth transition and fast ramp-up at the production site. A strategy that uses carry over parts (COP) helps a lot during that process because assemblers are already used to the parts at hand. For Airbus, it is difficult to follow such a strategy due to the fact that customers have very individual requests and their bargaining power is extremely high.

Airbus, for instance, ordered the auxiliary power unit (APU) for an aircraft as a single-source component. Some customers demanded the previous APU which they used as part of their carry over parts (COP) strategy. Besides bigger piece numbers leading to economies of scale it is also important for the airlines to consider extra efforts in professional training and documentation to implement a new APU as a standard component. The example shows how difficult it is for Airbus, compared with the automotive industry, to implement a carryover parts strategy. BMW, for example, identified 27 different product variants of air conditioners, years ago. They were able to reduce the total number to seven. The expenditure for construction, suppliers, assembly and documentation was reduced by a factor of 25. It was crucial that customers do not notice a reduction of product variants, which is why most changes were made behind the instrument panel. Methods and tools have to be further evaluated concerning transferability in order to exploit possible saving effects.

2.5 Change Management

Changes during product development or altering the product line is unavoidable but it is crucial to keep them at a minimum. Unfortunately, that is extremely difficult regardless of the industry. An aircraft consists of four million single components, it takes more than 10 years to develop it involving thousands of engineers and its product life cycle can reach 40 years. Naturally, the product changes continuously. Assuming a vehicle is containing 18,000 components, 1400–4000 change requests are made in the last 18 months of vehicle development before SOP. On average a change request refers to three components (Wildemann 2016c). Changes that concern the vehicles' design, for instance, radiator cowling, headlights, side panels or the luggage compartment mean significant extra effort. Even though these changes make only 10 % of the total amount they still account for most of the expenditure. The vast majority of the other 90 % concern technical alteration with regards to manufacturing feasibility, quality or they are based on test results. It is quite interesting that changes to the electric system or the equipment are not necessarily made before the SOP. Instead, they are implemented step by step with different facelifts. Almost 60 % of the required labor could be saved through process optimization.

Deducing from the automotive industry, one can imagine what kind of changes have to be applied during aircraft development. Changes in design are much more common due to the high bargaining power of customers. Some customers even deliver their own individual seats either themselves or by bringing in an additional supplier. Major elements such as kitchens or toilet units are placed individually, depending on the configuration that the customer choses. In consequence, changes and reinforcement also need to be applied to the structure individually. Low-cost airlines demand more seating capacity per aircraft. An A320 carrying 115 passengers has to have eight emergency exits in its body to be in line with security guidelines. With two additional rows of seats, it is obligatory to install two more emergency exits. That requires reinforcement of the body structure and further security certification. The depicted example shows how supplemental customer requests lead to higher costs and complexity. No matter if it comes to basic changes or detailed improvements, alterations have to be documented. The form of documentation should be standardized and include reasons, responsibilities, additional costs or necessary investments. In that way, guidelines can be created to stay in line with the company policy. Considering negative examples such as the A380's wiring harness or Takata's product recall for millions of cars due to broken airbags, both industries are in need of improvements. In both cases changes have not been properly communicated or coordinated. The experiences that have been made in both industries are very similar so that both parties can benefit from a knowledge transfer.

2.6 Production and Supply Chain

Aircraft production and supply chain differ greatly from those of an automobile. That is mainly because quantities, production processes, production time and the depth of added value are not alike at all. Technically, aircraft and automobiles are both made on demand. The main difference between each production and supply chain occurs at the point of the customer order specification. Car bodies are coated in different colors and versions based on statistics. Those versions are produced and then stored on a high bay warehouse where each body receives a specific identification number that can be linked to a customer's order. Suppliers receive new delivery orders accordingly and the assembly section is being informed about upcoming workload (Wildemann 2016d). In contrast, an airplane is built after a customer order is received. That also includes preparations on a structural level. The assembly, even though being a small part of the whole process, takes a lot of effort because of individual customer requests. The whole manufacturing process takes up to 1 year.

Another difference is associated with the supply chain. The automotive production following the just-in-sequence-concept is clocked down to the Tier 4 supplier with an ahead-of-schedule work of 4 weeks. Customers can therefore change color, motorization and equipment level up to 3 days before their car is produced. The whole process is based on efficiency which calls for suppliers that can react quickly. Changing the jet engine 3 days before SOP would cause a snowball effect. It would be necessary to change engine suspensions, steering, wiring harnesses and documentation which would be near the impossible. On the one hand both industries have in common that customer-specific design is extremely important. On the other hand the complexity of implementation differs greatly as well as the steering of the supply chain. Alongside resource purchasing and resource management, controlling of supplies and supplier structures (which includes communication and interface management) are the issues that can be beneficially transferred to Airbus.

2.7 *Continuous Improvement Process 2.0*

A new trend is causing excitement in automotive companies. It's called CIP through gamification. But what is that exactly and how can companies benefit from it? What is special about it? Generally CIP means that management constantly points out areas to act on and related potentials. Employees try to reach set targets while managing the whole process almost independently. The process can raise productivity by 3–5 % annually. Gamification relies on independent communication structures to solve problems using special programs. Gamification employs well known elements from games to make work more interesting, exciting but most of all more effective. For a lot of employees it is already a daily routine to play little games on their personal computer or mobile phone. The concept takes advantage of a habit and turns it into something productive. Doing so, it is not enough to just connect working with some kind of score. Immediate feedback, high relevance to the user and social components within the application are important features for successful programs. For a lot of companies, gamification still is an unknown concept. Applications have been used in marketing to promote consumer goods and to improve customer loyalty. Even though the concept's advantages and its practical approach are obvious, it has not yet been transferred to other divisions.

In the automotive industry, gamification has been used in the field of cost engineering (Wildemann 2016e). Cost engineering is the combination of customer value and optimizing costs during product and service development. The problem is to get employees to focus on the issue. Gamification should help with that. A communication platform connects important elements such as knowledge management, social networks and employees. The platform combines gamification and cost engineering. The economic necessity to reduce product costs stems from the fact that global markets demand different specifications, so products need to have local features in order to enter a new market successfully. That means two premises have to be kept in mind: over-engineering should be avoided and attributes have to be found that customers are willing to pay for. Cost-engineering measures in the automotive industry mostly follow a top-down approach. Incentives must be created to make cost-engineering a permanent issue for engineers and developers throughout the whole product life-cycle. Gamification is very suitable to get people to focus on such concepts. In addition to cost-engineering, the concept of gamification can be transferred very well and has a lot of potential value for Airbus.

2.8 Digitization and Visualization in Production

Automobile manufacturers are majorly affected by digitization. This does not only refer to digital manufacturing but also change processes, structures and the company's corporate philosophy. Industry 4.0 can possibly increase production by 9 % (Waltl and Wildemann 2014). Intelligent production facilities realize yet the slightest deviation which makes processes such as deformation of sheet metal very secure. Maintenance can also be controlled via the Internet which minimizes malfunctions. Other digital phenomena that will revolutionize the workspace are advanced robots that can assist human workforce, 3D-printers or driverless vehicles. A visible consequence of digitization is the growing number of employees specialized on information technology. The Volkswagen Corporation already employs 11,000 computer scientists, data analysts and software developers. Another 46,000 employees work in the research and development division. The working environment of the two groups is still different though. The digital world has a young vibe as companies promote entrepreneurial spirit and unconventional methods. A lot of results are achieved by using the trial and error method. Flat hierarchies are characteristic. Those companies just try out new ideas even though they are not fully developed which makes them agile and enables them to react quickly. On the opposite site, the automotive industry emphasizes established structures and processes. Just a dedicated car manufacturing process lives up to the standard of such a complex product and can ensure that everything goes according to plan. Mistakes such as those that are made in the information and communication industry must never be made in the automotive business. The vehicle has to function perfectly from day 1. That leads to the challenge of establishing new methods while still keeping well-proven processes and procedures.

The information and communication industry have functioned as a role model when introducing the smart factory concept because the two industries are used to dealing with technological change. Within the framework of the smart factory, a number of optimization projects were realized. The main fields which those projects concerned were automatization, networking, information management and ergonomics. The core of the projects is to connect previously independent physical or functional systems in order to create intelligent communication between humans, products and resources. Existing technologies were adapted and brought into the smart factory context. Cyber-physical systems are the basis for comprehensive networking in the smart factory (Waltl and Wildemann 2014). The concept includes machines and objects that use radio frequency identification chips or systems based on chip technology which enable to constantly send information about location, status or process progress. Scanners and computers can read, gather, forward and process the data. The key potential for success of the smart factory is the ability to manage big amounts of data (Waltl and Klein 2015). Process planning and controlling can be made automatic and autonomous. Intelligent production tools can use data effectively by independently reacting on deviations, material fluctuations and other influences because of their self-organizing skill. Data analysis helps to continuously monitor the process status. That includes remote machine maintenance via the Internet. Data simulations can be run to conduct virtual body construction or to test the SOP. Both can improve initial quality, output and expenditures. There are, for example, two different roofs for the Q7 that can be produced by using a fully-automatic, flexible framer (Waltl and Wildemann 2014). Automatization gives employees the opportunity to concentrate on monitoring tasks. Driverless transport systems which are currently tested as part of the Audi A6 production are another example for automated production. The transport devices are connected with each other and cover a distance up to 300 m. If a component part is needed, a system check is conducted to proof whether the route is available. After clearance, the transport devices start driving by themselves. The driverless system also sorts cars that have been ordered by the shipping department. The system Ray picks up the cars from the final production sequence and automatically puts them into shipping order. If radio frequency identification chips are put into the front bumper, the driverless transport system can automatically identify and process components. Audi also uses virtual and augmented reality applications. Computer-based representation combines the users' real and virtual world which can improve planning and development processes. One of these applications monitors robot movements during the body part construction. Another one helps to evaluate ergonomics and process security through virtual assembly planning. In the preproduction center, an automobile's body or power engine is represented on a project platform, to ease communication between project members. Tablets are used in the production lab to simplify engine compartment inspection. A barcode is used to display the engine and the necessary inspection steps in the right order on the tablet. Similar to the tablet concept, smart glasses can be used. With a camera which is attached to the glasses, the engine can be analyzed and recommendations can be made based on visual impressions. Reality and virtual representation are combined.

The smart factory concept also aims at improving collaboration with suppliers. Cyber-physical systems simplify supplies coordination. Production losses can therefore be minimized. The app Quick Check-in is already in use. It locates all commercial vehicles that are within a 50, 20 or 3 km radius. Geo-fence technology automatically sorts those vehicles by prioritizing shipments which can equalize delays. Just-in-time deliveries can arrive even more precisely. Cooperation between humans and machines is also key to the smart factory concept. Robots can release employees from routine works which makes it easier to deal with complexity (Köth 2015; Knight 2013). There are concepts such as 'production assistant hand parts' which aim at improving ergonomics at the work station. During the expansion tank assembly process the object is directly handed over to the worker. The chairless-chair also improves ergonomics by supporting the employee. It is an exoskeleton supporting the employee during assembly. Joints are relived because a worker can always sit down and assemble pieces in a healthy body position. Considering demographic changes, these concepts are especially important because they can guarantee that older employees can carry out their tasks without any restrictions. The luggage compartment door assembly is also supported by robots. A robot measures the luggage compartment door with its sensors and compares the data to the body part parameters. Based on the result, it positions the luggage compartment door. The worker is solely in charge of the screw connection. Luggage compartment doors are very big and heavy, they have sharp corners hence security measures must be taken. Laser scanners monitor both robot and worker movements at the work station. In case of a possible collision, the robot's movement is immediately interrupted to avoid any kind of danger. Humans will still play an important role in the smart factory because of their experience and intelligence. The results that have been gathered in the automotive industry are highly relevant for Airbus. They can be transferred extremely well to aircraft production.

2.9 Optimization During Assembly

The assembly of a VW Polo took around 21 h on average in 2005 (Harbour 2013). Since then, the number of used pieces increased by 12 %. What optimization process reduced the assembly time by 16 %? Mainly responsible were a product design with focus on easy assembly, new manufacturing technologies, innovative work systems for process design and uncountable CIP measures in order to optimize the assembly process. The product design accounted with 4 %. CIP measures and work systems accounted for 3 % each. The other 2 % were reached by using new manufacturing technologies (Waltl and Wildemann 2014). The areas of influence to improve production are CIP measures to optimize assembly, the assembly line sequence and the use of new manufacturing technologies (Wildemann 2011). Great manufacturing technology improvements could be made by introducing joining technologies that use clips and by introducing manipulators that could help moving heavy loads. Manipulators are auxiliary devices that improve ergonomics and enable a worker to assemble heavy, bulky component parts such as dashboards, seats, doors and window panes. Assembly process optimizations are realized by using applicable trays and improved route management following the "Werkerdreieck" (Wildemann 2016f). The term refers to three sequences: component reception, installation and return to the starting point.

Using the example of the e-Golf, what do these concepts mean? The e-Golf is a derivative of the product portfolio and is based on the combustion engine model. Sales quantities are uncertain, therefore both models have to be manufactured on the same production line. The biggest difference stems from the two unequal construction types which requires substitution of the combustion engine, gearbox, exhaust tract, heat conduction plates and fuel container. Those parts are replaced by an electric engine including the gearbox, a battery, a charging device including a high-voltage supply system on board, power electronics, heating and a brake booster. Complexity is caused by missing experience with these new technologies in product development and assembly. Besides construction, the focus during product development lies on alignment of hardware and software to assure functionality. For a mixed manufacturing process it would be necessary to match assembly, tool design and construction, logistics and most of all employee qualification due to the different vehicle architecture. In the pilot construction center a concept must be found and be proven that integrates the new drive concepts manufacturability with the established Golf design during the early stages of product development. The testings' emphasis lies on a digital test model, constructed space audit and ergonomic assembly including wiring and attachment points as well as accessibility in case of repairs. Furthermore, assembly sequences and knowledge management are important in order to reduce knowledge gaps. The virtual Golf model is constructed when its concept is determined 2 years before SOP. After the digital model is cleared, an e-Golf prototype is constructed in the prefabrication center. Besides adding missing or defective data, technical features are evaluated. The project management of the prefabrication center gives assignments to the assembly section as well as it orders materials, special tools and devices based on the computer-aided design. The aim is to do a real-time check on feasibility and manufacturability. Building the first prototypes is the first step of analyzing and testing the series production. A major focus is placed on conventional mistakes, assembly process evaluation in consideration of production time, functionality and dimensions. That is realized by repeatedly assembling and disassembling parts and components using manipulators and the "Werkerdreieck". Afterwards the prototypes are evaluated and it is reflected on possible changes in the construction. Next to product and part adjustments, an emphasis lies on adapting procedures. It is of importance to apply changes to secure the high-voltage system for logistics, assembly and initial operations. Employees have to be familiarized with new testing equipment and labels. Constant quality assurance in all areas facilitates small control loops which benefits all project members. The loop is repeated until the series receives clearance. Interface management and performance agreements are just as important as costs, deadlines and qualities with regards to the SOP of the Golf. In most cases, this task is carried out by an experienced ramp-up project team that is selected depending on the vehicle. Frequent reporting including an escalation strategy is indispensable. Car development and independent vehicle production heightens complexity because coordination efforts and influence on the concept by the preproduction center increase disproportionately. The reasons for that are additional technological changes in areas such as press shop, body construction and paint shop.

Based on an approach that aims at creating solutions a lot of advantages may unfold. Challenges can be detected early during the vehicle project. It is easier to stick to the plan during the preproduction phase concerning all model series considering planning and production requirements, close networking between development, testing plus planning and production. Finally, assembly processes can be identified early and then be optimized and secured which enables a quick and smooth transition until the desired capacity is reached. Judging from different vehicle projects, the preproduction center has proven to be the best practice and knowledge platform. It is essential for employees to receive professional training in the preproduction center academy. Only if transparency, openness, training programs and advanced future competency areas are promoted, increasing product variety and ramp-up can be managed successfully by the preproduction center. It is responsible for the series' product and production requirements on a virtual and physical level, from product definition until the desired production capacity is reached. Exchange between the centers of different product models and brands enable a faster, cheaper and higher-quality ramp-up for new vehicles. This example should show, how important it is in the automotive industry to focus on different concept interaction. At the same time it is the basis for discussion about different applications at Airbus. Figure 2 provides an overview of the concepts presented as well as it shows how well which concept can be transferred.

3 Summary

It needs to be noted that similarities exist despite great differences in product and production. Most of all those are challenges in the fields of customer individualization, number of modifications during the development process, the general development process, efficient production and logistics processes, a big number of carry over parts, modularization of assembly groups and constructions, the supply chain management and increasing value chain digitization. The concepts presented such as multi-product and multi-brand plants, production segmentation, standardization and modularization, ramp-up and change management, optimization of logistics and commissioning, continuous improvement process 2.0 as well as digitization and visualization of product and production optimization during the

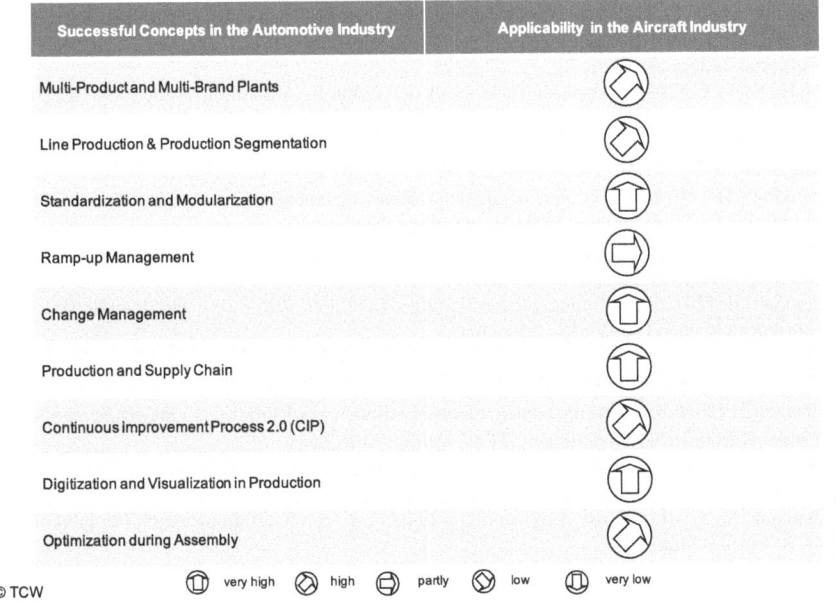

Fig. 2 Applicability of the most successful concepts in the automotive industry for the aircraft industry

product development process as well as assembly optimization play a great role in order to overcome those challenges. Reflection on both sides can lead to great room for improvement in both industries.

References

Alcock C (2015) Airbus to acquire flight support group Navtech. Aviation International News, Edition Dec 2015. Available via http://www.ainonline.com/aviation-news/air-transport/2015-12-22/airbus-acquire-flight-support-group-navtech. Accessed 11 Apr 2016

Harbour R (2013) The Harbour report 2013. Oliver Wyman, Berkley

Hojak F (2015) Gestaltung des Restrukturierungscontrollings—eine theoretische Betrachtung und empirische Untersuchung. TCW-Verlag, München

Knight W (2013) Smart robots can now work right next to auto workers. MIT Technology Review (09/2013). Technology Review. Available via https://www.technologyreview.com/s/518661/smart-robots-can-now-work-right-next-to-auto-workers/. Accessed 11 Apr 2016

Köth CP (2015) Roboter erobern die Montage. Automobil Industrie (10/2015): 58–59

Meißner HR (2013) Die Bedeutung der Automobilindustrie für die deutsche und europäische Wirtschaft. Agora42 (03/2013). Available via http://www.blicklog.com/2013/10/31/die-bedeutung-der-automobilindustrie-fr-die-deutsche-und-europische-wirtschaft/. Accessed 8 Dec 2015

Munde A (2015) Smart produzieren. Automobil Industrie Sonderausgabe Produktion 2015:12–14

VDA-Arbeitskreis PLM (2012) Product-lifecycle-management. Empfehlung 4961/3, Ausgabe April 2012. Available via https://www.vda.de/dam/vda/.../1333357251_de_1291783044.pdf. Accessed 22 Dec 2015

Waltl H, Klein C (2015) Daten sind das Gold der Zukunft. Automobil Produktion (09/2015): 78–81

Waltl H, Wildemann H (2014) Modularisierung in der Automobilindustrie. TCW-Verlag, München

Wildemann H (1997) Fertigungsstrategien—Reorganisationskonzepte für eine schlanke Produktion und Zulieferung. TCW-Verlag, München

Wildemann H (2011) Neues Montagemanagement—Neue Montagekonzepte in der Kleinserienmontage komplexer Produkte. TCW-Verlag, München

Wildemann H (2016a) Fertigungssegmentierung—Leitfaden zur fluss- und logistikgerechten Fabrikgestaltung. TCW-Verlag, München

Wildemann H (2016b) Anlaufmanagement—Leitfaden zur Verkürzung der Hochlaufzeit und Optimierung der Anlaufphase und Auslaufphase von Produkten. TCW-Verlag, München

Wildemann H (2016c) Änderungsmanagement—Leitfaden zur Einführung eins effizienten Managements technischer Änderungen. TCW-Verlag, München

Wildemann H (2016d) Supply Chain Management—Leitfaden für ein unternehmensübergreifendes Wertschöpfungsmanagement. TCW-Verlag, München

Wildemann H (2016e) Cost Engineering—Leitfaden zur Gestaltung von Produktkosten. TCW-Verlag, München

Wildemann H (2016f) Montagemanagement—Lösungen zum Montieren am Standort Deutschland. TCW-Verlag, München

Horst Wildemann, Univ.-Prof. Dr. Dr. h.c. mult., is Head of the Research Institute for Corporate Management, Logistics and Production and Managing Partner of TCW Management Consulting. His practical experience has been the foundation for his 40 books and more than 700 essays in which he shows new ways to create a profitable successful company. Prof. Wildemann works as a consultant, as a member of the board of directors and as an advisory board member for leading industrial enterprises.

Florian Hojak, Dr., Head of Automotive and Manufacturing of TCW Management Consulting. Before working with TCW, he had been a manager in an automotive OEM consulting firm for several years and in an international restructuring advisory services firm.

Part III
Component Manufacturing

Trends in the Commercial Aerospace Industry

Gernot Strube, Karel Eloot, Nadine Griessmann, Rajat Dhawan, and Sree Ramaswamy

Abstract McKinsey's prior work on the future of manufacturing discusses five broad trends that together are shaping the industries that make up the global manufacturing sector, ranging from automotive to textiles. These five trends are (i) demand growth, (ii) supply costs, (iii) business risk, (iv) technology and innovation, and (v) policy and regulation. Commercial aerospace, a roughly $300 billion global industry, is also affected by these five trends. Taken together, the trends presage a gradual shift of the global aerospace manufacturing footprint toward the Asia-Pacific region. The shift will serve rapidly growing local demand and respond to local government action that seeks to exploit a cyclical upturn in demand to attract investment in local manufacturing. Supply chains are already expanding in the region as Western incumbents build up capacity; as one or two new OEMs emerge from this region, particularly from China, local supply chains are likely to become deeper in capability as well. Given the long time investment (5–7 years) to build up capacity, the 'nationally important' status of the industry, and the risks that incumbents face in intellectual property, quality certification, and collaboration in both product development and manufacturing, these shifts in the aerospace footprint may continue to lag the footprint shifts of automotive and other manufacturing industries. But the shift to Asia-Pacific, already accelerating in the past decade, could speed up even more under two or three scenarios related to increased competition, pressure on profit margins, and expansion of digital capabilities in the aerospace value chain.

G. Strube (✉) · N. Griessmann
McKinsey & Company, Munich, Germany
e-mail: gernot_strube@mckinsey.com

K. Eloot
McKinsey & Company, Shanghai, China

R. Dhawan
McKinsey & Company, New Delhi, India

S. Ramaswamy
McKinsey & Company, Washington, DC, USA

1 Introduction

The global commercial aerospace industry generated roughly $300 billion in annual revenues in 2015 (McKinsey Global Institute and Aviation Week 2015). Manufacturing activities accounted for nearly $240 billion across OEMs and suppliers of engines, aerostructures, avionics, and other subsystems. Aircraft OEMs accounted for slightly over half of this revenue; of the rest, engines accounted for one-third. In addition to revenues associated with manufacturing activity, the aerospace industry also generated $60 billion in revenues in aircraft maintenance, repair, and overhaul (MRO) activities worldwide. Although the industry is global from a sales perspective, it is less so from a production perspective; North America and Europe, for instance, accounted for about half of the global commercial aircraft deliveries in 2015, but the two regions represent 90 % of total production, and aerospace firms from the two regions make up 95 % of total sales (McKinsey & Company's aerospace and defense practice). MRO activity, on the other hand, benefits from localization and therefore tends to more closely reflect the geographic dispersion of aircraft fleets.

As the global economy recovers from the 2008 financial crisis and recession, five broad trends are shaping the global manufacturing sector and affecting industries as varied as automobiles, metals, pharmaceuticals, and textiles (McKinsey Global Institute and McKinsey Operations Practice 2012). Manufacturing companies face long-range shifts in their environment as a result of these five trends—including shifts in patterns of demand, rapid changes in factor input costs, the spread of new technologies and innovations, proactive government actions to foster and support domestic production, and new forms of business risk, including new global competitors and threats to future returns on capital. Commercial aerospace is no exception to these trends.

2 Demand

The commercial aerospace industry is benefiting from an ongoing cyclical upturn that is expected to continue, though recent global economic uncertainties could slow its progress. Global aircraft fleets are forecast to more than double over the next 20 years, propelled by strong demand growth in emerging markets (Fig. 1) and improved financial conditions for airlines. Europe and North America are not likely to remain the world's largest commercial aviation markets. Sixty percent of new deliveries are to markets outside Europe and North America, and the Asia-Pacific region is likely to become the world's largest commercial aviation market. The demand growth has generated industry records in recent years for both deliveries and backlogs. Demand for aftermarket MRO services is increasing rapidly as well, driving a greater need for localized sales and service activities.

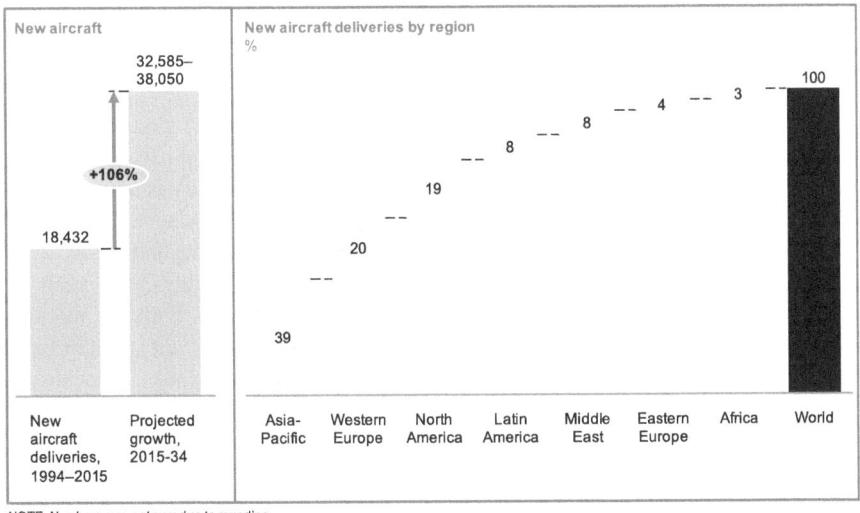

Fig. 1 Commercial aircraft deliveries are forecast to more than double over the next 20 years compared to the past 20 years. Source: Airbus Global Market Forecast, 2015–2034; Boeing Current Market Outlook, 2015–2034; McKinsey Global Institute Analysis

2.1 Commercial Aircraft Demand Is Shifting Rapidly to Emerging Markets

It is widely known that economic growth has shifted toward developing economies, but the momentum of that shift is not fully appreciated. Total consumption in emerging markets could rise from $12 trillion annually in 2010 to $30 trillion in 2025 (McKinsey & Company 2012). As developing economies grow wealthier, some 1.8 billion individuals are likely to enter the global consuming class, and 60 % of households in the world with incomes of at least $20,000 a year will likely be in developing economies (McKinsey Global Institute 2013). By 2025, emerging markets could account for nearly 70 % of global demand for manufactured goods (IHS Economics).

The magnitude and speed of this shift is playing out in commercial aerospace as well, as rising household incomes are expanding the class of consumers who can afford air travel. On a per capita basis, by 2034 the people of China will be flying as much as Europeans today, according to Airbus (Airbus GMF 2015–2034). In addition, airline companies are generally in a stronger financial position, more confident about absorbing risks such as fuel price fluctuations and financing costs. Global aircraft fleets are forecast to more than double over the next 20 years, propelled by strong demand growth in emerging markets, particularly for narrow-body aircraft. Roughly 60 % of deliveries are to markets outside Europe and North America, and the Asia-Pacific region is expected to become the world's largest market for commercial aviation, with ten of the top 20 traffic flows.

2.2 Aerospace Is Less Exposed than Other Industries to the Need for Market Proximity

In most manufacturing industries, these massive shifts in demand would presage a corresponding shift in production as well. Roughly two-thirds of global manufacturing value added is in industries where market proximity is an important factor in location decisions (George et al 2013). They include capital goods industries such as machinery and automobiles, regional processing industries such as food and beverages, and energy- and resource-intensive commodities such as some metals and plastics. In automobiles, for instance, China is both the world's largest market and its largest producer; it is also home to a significant supplier base that has followed many Western automotive OEMs as they look to locate in proximity to the booming Chinese market. Furthermore, markets in Africa, Brazil, China, and India are not monolithic—they are made up of extraordinarily diverse regional, ethnic, income, and cultural segments, most of which can be large enough to compare to entire developed-nation markets. For example, at roughly $600 billion, Shanghai's GDP (in terms of purchasing power parity) is larger than that of Belgium, Switzerland, and Norway. The proliferation of markets, as well as rising requirements for customized products, is fragmenting demand; companies need to produce more local-market variations and ship a wider variety of stock-keeping units to compete.

Commercial aerospace is an exception to the trend of demand proximity. As they look to expand their production network, industry incumbents remain wary about concerns such as optimizing their footprint in a low-volume, high-complexity environment with long production and life cycles, intellectual property protection, productivity differences, considerations of national interest, and certification challenges. As a result, supply chains have been slower to migrate near markets for aerospace compared to industries such as automotive. While Airbus recently extended its Tianjin final assembly contract for another decade, for instance, most components and subassemblies are still aggregated in Hamburg before being shipped to Tianjin. In 2014, for instance, Airbus was building four or five aircraft per month in Tianjin, around 10 % of its monthly global volume of A320 aircraft (Airbus 2014).

2.3 Demand for Aftermarket Services Continues to Grow

Finally, another noteworthy change in demand patterns for manufacturers in certain sectors—particularly those that sell to other businesses (business-to-business, or B2B, segments)—is the growing demand for value-added services, and software to go along with manufactured goods. This has raised the share of manufacturing sector revenue and employment associated with services to as high as 55 % in some manufacturing industries (McKinsey Global Institute). The demand for aftermarket services is highest in capital goods industries. For example, in electrical and industrial machinery, services account for 30–40 % of the total cost of ownership. In transportation equipment, such as fleet vehicles and forklift trucks, services can represent as much as 40–45 % of the total cost of ownership. In comparison, services make up less than 10 % of the total cost of ownership for commodity manufactured goods such as appliances, furniture, and commodity chemicals.

Demand for aftermarket services is strong in commercial aerospace. Already many segments of the aerospace value chain derive 40–50 % of their revenues from the aftermarket through MRO activities; MRO margins, especially for spare parts, also tend to be higher than on direct sales of original equipment. An additional benefit to aerospace companies is that MRO activity tends to be less cyclical than aircraft sales. Aerospace firms provide a growing number of pre- and post-sale services to their customers: maintenance, logistics, inventory services, leasing, financing, risk sharing, and training and support. Private defense companies, for example, increasingly provide leased aviation services, including pilots, air-to-air refueling, and "power by the hour." In developed markets, there is considerable competition between aircraft manufacturers, equipment suppliers, and traditional service providers (such as airlines and third-party aviation service providers) to determine who captures the value. As demand for commercial aircraft expands in emerging markets, demand for aftermarket services is also growing and is driving the need for localized sales and service activities.

3 Supply

Despite the rapid growth of demand in emerging markets, the aerospace production footprint remains primarily in developed economies. Over 80 % of today's value added in the sector comes from traditional production centers in North America and Europe. In these regions, an aging workforce and shortage of engineering talent are the primary supply-side concerns for manufacturers. More than a quarter of aerospace workers could retire in the next decade. The commercial aerospace industry continues to lag other manufacturing industries in offshoring value-added activities to lower-cost locations. But new, lower-cost locations will continue to grow, especially as a source of components and subassemblies as emerging markets attract investment based on a range of strengths. China's value-added share in aerospace and defense has doubled to 8 % in just the past 5 years. Lower labor costs and a younger workforce make emerging markets attractive, and the importance of China, Eastern Europe, Mexico, and India as sourcing partners is growing—but wages are rising rapidly even in low-cost locations, creating an incentive to accelerate labor productivity growth.

3.1 New Locations Will Continue to Grow as Sources for Global Aerospace

Today, 80% of the commercial and defense aerospace industry's value added comes from Western Europe and North America, down from 90% in 2000. The Asia-Pacific region accounts for 14% of value added, up from 5% in 2000—with China doubling its share of global value added to 8% in just the past 5 years (Fig. 2). Other countries in the region, such as Japan, Korea, and Singapore also make up a significant share of global value added. The shift to Asia-Pacific is likely to continue over the next decade as this region becomes the world's largest commercial aerospace market. In 2015 Boeing announced the establishment of a completion and delivery center in China for its 737 aircraft (Boeing 2015). In many industries, the shift of demand is also changing the nature and pricing of products that manufacturers must sell in order to stay competitive. In commercial aerospace, for instance, aircraft OEMs are facing pressure from airlines that demand more flexibility in customizing configurations to meet the needs of specific routes. For OEMs, this translates to a demand for more customized configurations. In 2014, Airbus announced it was exploring options to develop a wide-body aircraft completion center in China (Airbus 2014).

China is the largest player among emerging markets in the commercial aerospace industry due to a combination of factors: a large and growing market, major infrastructure investment, a relatively low-cost labor pool, and government focus on expanding the domestic aerospace industry. But other emerging markets in Asia, Eastern Europe, and Latin America are also becoming sources of aerospace supply, and they have their own advantages. Mexico, for instance, is one of the world's largest recipients of foreign direct investment in aerospace in recent years. Among other attributes, the country benefits from its proximity to North American OEMs,

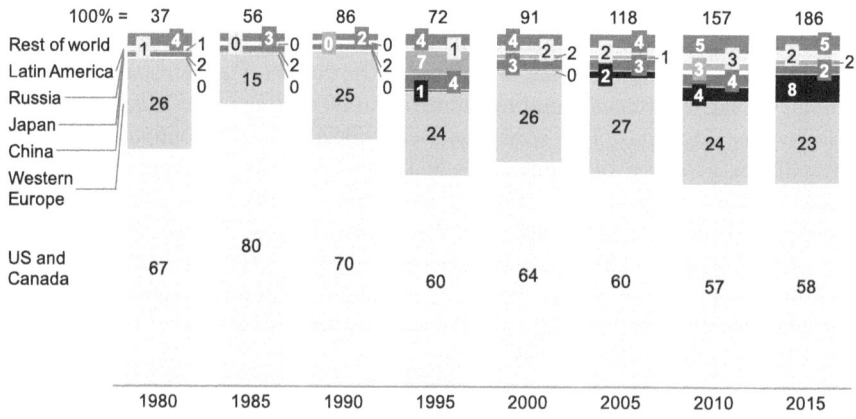

Fig. 2 North America and Western Europe account for 80% of aerospace value added; China's share is rapidly growing. Source: IHS; McKinsey Global Institute Analysis (2015)

its technical skill base, and its trade connectivity with other fast-growing regions. India shares some of China's advantages, especially its large market, relatively low-cost labor pool, and nationally important aerospace and defense companies. Although the country has not committed to developing a homegrown OEM, the aerospace sector is one of India's priorities as it seeks to become a global hub for manufacturing.

3.2 Talent Shortages Are Especially Critical for the Aerospace Industry

In a 2015 survey, 32 % of employers from European, Middle Eastern, and African nations and 83 % of Japanese companies reported having difficulty filling jobs because of a lack of qualified talent, particularly technicians, skilled trade workers, and engineers (Manpower Group Talent Shortage Survey 2015). Past MGI research has highlighted potential shortages of high-skilled workers around the world and potential oversupplies of less skilled workers. Three of the top ten hardest-to-fill jobs in 2015—technicians, skilled trades workers, and engineers—are directly relevant to knowledge-intensive manufacturing industries such as aerospace. Developing economies may not be able to address these shortages; in Brazil, China, and India, the rapid growth of knowledge-intensive manufacturing is expected to create shortages of both high-skilled engineers and middle-skilled technicians by 2030.

The global aerospace industry is especially affected by talent shortages, because it also faces a demographic challenge due to an aging and slower-growing workforce. In the next two decades, the growth of the global labor force will slow; in many advanced economies, the growth will average less than 1 %. The aerospace industry is more exposed than most other industries. The median age of workers in the US economy is 42 years, but in the aerospace industry it is nearly 48 (United States Bureau of Labor Statistics). More than 25 % of US aerospace industry workers are over the age of 55, and many could retire in the next decade; including workers in the 45–55 age group swells the share of workers to nearly 55 %. Meanwhile a disproportionately high share of middle-tenure technical employees have left the aerospace industry, creating an age gap in the workforce and challenging succession plans. Many firms also report challenges in recruiting, developing, and retaining top talent.

3.3 Wages Are Rising Rapidly in 'Low-Cost' Locations

As developing economies continue to industrialize, rising wealth and productivity are also associated with rapidly rising wages, often due to specific policy actions, as in the case of China. From 2006 to 2013, real wages in the group of advanced economies grew at no more than 1 % per year (International Labor Organization). In those years, real wages in Asia grew by 5.1–7.7 % annually, and nearly twice as fast in some emerging Central and Eastern European countries. Meanwhile, industrial automation continues to make inroads, changing the labor-capital tradeoff dynamic. In the past two decades, the cost of automation has fallen by 50 % or more relative to labor in several developed economies.

Rising labor costs in developing economies, and fewer opportunities for labor arbitrage thanks to declining automation costs, could make it even less likely that wage differentials will play a role in determining the future aerospace manufacturing and supply chain footprint. In general, rising wages in low-cost locations affect manufacturers only when the trend materially changes the total landed costs of production. This is most likely in labor-intensive activities such as apparel manufacturing and electronics assembly—places where labor is a relatively large fraction of compressible costs. In these industries, some companies respond to rising wages by moving to lower-cost locations, but only if the footprint is relatively 'un-sticky'. For many electronics assemblers, for instance, proximity to the East Asian supply chain trumps any labor cost advantages in other parts of the world.

4 Business Risk

The rapid growth of demand in emerging markets is spurring the rise of new competitors to the large Western OEMs and suppliers. While Western firms remain the dominant suppliers, accounting for 80 % of industry value added, these new rivals are a potential competitive risk. This is especially true in the fast-growing Asia-Pacific market, where it is crucial for the industry to grow revenue and maintain return on capital. As OEMs and tier-1s look to expand in emerging markets to meet demand and compete against newer rivals, low-cost sourcing could become more critical, particularly in the more fragmented parts of the value chain, where profit margins also tend to be lower. But challenges around IP protection, technical capability, and build quality remain. Finally, supply chain complexity remains a significant business risk in the aerospace industry. Recent trends toward more outsourcing, greater transparency, and focus on partnerships mean that risk is being shared between aerospace companies and their suppliers regardless of their tier.

4.1 New Competitors Are Emerging from the Asia-Pacific Region

Over the past two decades, rapid growth in emerging markets, particularly in capital-intensive sectors such as infrastructure, transportation, capital goods, and natural resources, has fueled the dramatic rise of home-grown competitors from these countries. Since 2000, the share of developing-economy companies in the Fortune Global 500 has grown from less than 5 % to nearly 30 %; if the trend continues, these firms will make up nearly half of the world's largest companies by 2025. They have already changed the rules of global competition in many of the capital-intensive industries in which they operate, by exploiting a lower cost base and fewer legacy assets to aggressively invest in capacity and capture market share. Their rapid growth and capital turnover allow these firms to generate returns on capital even at the expense of profit margin. Many of these firms are state- or family-controlled, allowing them to take a much longer perspective and focus on growth and scale instead of short-term returns. As a result, while Western corporate profits are at record highs, the composition has changed dramatically since 2000; profits are increasingly accruing to asset-light, intangible-heavy firms and industries while incumbents in capital-intensive sectors, facing these new competitors, struggle to break away from the pack.

These trends have not affected the commercial aerospace sector yet, but new competitors are on the horizon. The most likely source of these new firms is China, home to a narrow-body competitor (COMAC's C919) that is expected to enter service shortly. Supply chain competition could also come from Japanese firms that have grown their share of the aerospace supply chain since the mid-1980s, when the Aircraft Industry Promotion Law was revised to encourage domestic firms to participate in international collaborations. Japan's value added in global aerospace has grown at 7% per year in real terms since then. For Western incumbents, these new challengers are a competitive risk, especially in the fast-growing Asia-Pacific region. Roughly 15% of aircraft deliveries in the next two decades are forecast to go to China, so even a limited local market entry could allow COMAC and its suppliers to quickly achieve market share and accelerate the transfer of value-added share to China.

4.2 Low-Cost Sourcing May Become More Critical but Remains a Challenge

As OEMs and tier-1s look to expand in emerging markets to meet demand and exploit lower costs, they must overcome challenges around IP protection, technical capability, and build quality. For North American and European incumbents—both OEMs and suppliers—the rise of new competitors poses a tricky challenge. Many of these firms have already established joint ventures in China, some for engines, electrical systems, and control systems. Chinese firms have been suppliers for airframe OEMs and tier-1 suppliers, but the growth of a lower-cost Chinese competitor may force incumbents to reevaluate their business strategy in some emerging markets. Aerostructure suppliers, for instance, have among the lowest operating profit margins in the aerospace value chain and are among the most fragmented; in other industries, similar conditions have seen the rise and consolidation of emerging-market suppliers that benefit from a lower cost base, higher economies of scale, and long-term ownership structures. In aerospace, recent trends suggest a shift to greater transparency and partnership in supply chain sourcing.

With more transparent costs and pricing, there may be greater opportunities for low-cost suppliers to break into the supply chain.

The emergence of a Chinese competitor could also strengthen the domestic supply base, allowing incumbent manufacturers more options in sourcing. Western aircraft OEMs have been sourcing from China for several decades, starting in the mid-1970s with low-complexity parts (such as nose sections, doors, stabilizers, and fins). Over time they have sourced larger structures, and over the past decade Chinese suppliers have signed risk-sharing agreements, provided single-source supplies, and participated in joint ventures with large Western OEMs and tier-1s, albeit still at a low level compared to other sourcing destinations. Despite this growth in low-cost sourcing, the commercial aerospace industry continues to lag most other manufacturing industries in terms of offshoring value-added activities, especially to lower-cost locations. North America and Europe make up more than 80 % of global value added in aerospace manufacturing, but they make up less than 50 % of global value added in manufacturing overall. The potential for low-cost sourcing remains high; a landed-cost analysis of aircraft structure production, for instance, shows that costs could be lower by 20 % in low-labor-cost locations. But aerospace manufacturers remain wary of intellectual property leakage, poor transport and utility infrastructure, and the lack of system integration and managerial talent to handle large programs.

4.3 Supply Chain Complexity Poses a Formidable Challenge

The manufacturing sector comprises a range of industries that differ greatly in the size and complexity of their supply chains. Some manufacturing industries, like food and pharmaceuticals, have relatively short supply chains; others, like automobiles, have long ones. The aerospace supply chain is the longest and most complex of all. This becomes a factor when large commercial aircraft OEMs and their suppliers must deliver significant production ramp-ups over the next decade to meet rapid growth in demand. The primary challenge is to ramp up production of aircraft that are being offered with new, more fuel-efficient engines, such as A320neo and 737MAX. And while the volumes of aircraft such as the 787, the A350 and the 777X are lower, the complexity of these aircraft greatly exceeds that of narrow-body aircraft.

Production ramp-up could become a business risk, past trends indicate; over the past 10–15 years, for instance, the commercial aerospace industry has suffered repeated and well-publicized delays and cost overruns both in initial deliveries and in subsequent production ramp-ups. A wide range of issues caused the delays, highlighting the extraordinary complexity of the aerospace supply chain. Over the next decade, successfully ramping up output presents additional challenges to OEMs and their suppliers, as new technologies are being brought to market and new capital investments are needed to expand capacity. The length and complexity of aerospace supply chains similarly makes managing ongoing cost reduction initiatives more difficult.

These challenges also highlight the increasing reliance of OEMs on their suppliers, as OEMs have outsourced an increasing proportion of their production. Operating profit margins tend to be significantly higher for aerospace suppliers with high IP content—particularly in the more consolidated parts of the supply chain, such as engines, avionics, and aircraft systems—compared to OEMs. In the past decade, both Boeing and Airbus embraced a model of transferring more activity, and risk to a smaller set of large-scale partners who could share capital expenditure, development costs, and risk. Meanwhile, tier- 2 and tier-3 suppliers may continue to see a greater push for consolidation, particularly in the more fragmented segments of the value chain such as aerostructures and aircraft systems. While there are benefits to having this more "centralized" outsourced model, the jury is still out; gains in program management and focus on higher-value-added activities are offset by the risk of increased complexity and integration challenges.

5 Technology

In the technology domain, key trends affecting the commercial aerospace industry range from advances in materials technology to additive manufacturing and digital technologies. The share of composites in aircraft has grown steadily, to 50 % of structural weight in the 787 and the A350. But as demand for composites ramps up in industries such as automobiles and wind energy, that could pose an availability risk for aerospace manufacturers, additive manufacturing continues to gain traction. Use in small-batch, high-complexity applications could transform some parts of the aftermarket supply chain, but deployment at scale is likely to be slowed by issues of accuracy, cost, speed, and qualification for critical components. Meanwhile, digitization holds significant promise for productivity improvements. The industry may see the potential expansion of a Digital Thread beyond supply chain management to improve production workflow, reduce equipment downtime, and represent a greater share of value creation (together with data analytics) in the final product. Other technologies such as commercial unmanned aircraft, while potentially disruptive, require significant regulatory and technical solutions—and in any case, their first commercial use could more likely be for low-end, short-haul cargo than for long-distance air transport of passengers or freight.

5.1 Aerospace Manufacturers Could Face Supply Challenges for New Materials

High-strength steel, aluminum, and carbon composites have been an important part of industrial design and manufacturing since the 1970s. Today the drive for resource efficiency and carbon emissions reduction is driving more widespread use. For instance, carbon fiber composites accounted for 5 % of aircraft design in the 1980s; in the 787 Dreamliner, composites account for 50 % of the plane's weight. The use of new materials in aerospace continues to grow, though technical challenges persist, notably in changing traditional production techniques such as stamping and spot welding, improving cycle times for composite materials in automated production lines, ensuring that computer simulations can accurately model material stresses, and finding ways to recycle synthetic materials such as thermoset resins.

But even as commercial aerospace manufacturers look to ramp up the use of composite materials, they may face a supply challenge from other industries that are also increasing the demand for composites—notably the automobile and wind-turbine manufacturing industries. Many German and Japanese carmakers are partnering with carbon fiber suppliers to use the material more widely as a way to reduce vehicle weight and improve fuel efficiency without sacrificing safety and structural integrity. As carmakers expand production of electric vehicles, the demand for composite materials could accelerate as manufacturers try to compensate for increased battery weight. Meanwhile, as the renewables industry continues to expand globally, total installed wind capacity worldwide could more than triple by 2030 (Global Wind Energy Council 2015).

Under current trends, demand for carbon fiber could reach 600 kt—about 20 times the current demand—causing bottlenecks in the aerospace supply chain and competition for resources with industries such as automotive. But the supply chain risk is not limited to carbon fiber composites. To take one example, the aerospace industry's move to carbon structures requires titanium to replace aluminum for adjacent structures to avoid corrosion. That substantially increases the demand for titanium. The automotive industry's move to lightweight materials and new power train and chassis technologies can put significant strain on the supply of aluminum, carbon, and rare earth materials. It is estimated that a shift to electric drive trains in autos could raise demand by carmakers for rare earth materials such as neodymium from 15 % of current global production to 550 % within the next decade.

5.2 Additive Manufacturing Continues to Make Inroads in Complex Applications

For the commercial aerospace industry, additive manufacturing systems continue to hold out significant promise. Manufacturers could see benefits in component design, with previously challenging designs now possible with no additional cost to complexity. Tooling costs could be significantly reduced, and tooling lead times almost eliminated. Material use and waste could also be reduced; Airbus Group research found, for instance, that titanium powder can be used to print components that are as strong as machined parts and use only 10 % of the material—important to OEMs that are becoming concerned about the security of the titanium supply. Finally, additive manufacturing could significantly improve operational flexibility, allowing manufacturers to simplify supply chains, reduce inventory (especially for spare parts), and set up production near MRO facilities.

Yet achieving these benefits at mass scale is likely to take years rather than months (Cohen et al 2015). Of the roughly $70 billion global market for machine tools, additive manufacturing systems represent only a fraction, despite double-digit growth in some types of printers. The cost of printing remains high, both in terms of the machines and maintenance (40–60 % of total printing costs) and in terms of materials. Costs may fall over the next decade as patent expirations and new Asian suppliers put downward pressure on prices and as economies of scale kick in for powder suppliers. Technical challenges in speed, accuracy, and throughput remain. Operational challenges may be an even higher hurdle, ranging from qualifying 3D-printed components to reorienting supply chains. But the aerospace industry remains one of the foremost adopters of additive manufacturing for complex parts and components.

5.3 Aerospace Supply Chains Can Benefit from the Digital Thread

Digital technologies have transformed many industries in the past 30 years, particularly retail, financial services, media, and other consumer-facing sectors. More recently, companies in asset-intensive industries such as mining, manufacturing, and utilities have been stepping up digital investment. Rapid and disruptive growth in computing capabilities, data generation, and data storage, along with advances in other areas such as advanced analytics, automation, human-machine interaction, social technology, and machine-machine communication are ushering in a digital revolution in the global manufacturing sector (Hartmann et al 2015). Leading manufacturers are in the early stages of connecting their physical assets using a Digital Thread—a seamless flow of data across the value chain to link every phase of the product life cycle, from design, sourcing, testing, and production to distribution, point of sale, and final use (Fig. 3).

Analyzing the after-sales data reported by sensors embedded in complex products enables manufacturers of goods from aircraft to data center servers to refine preventive maintenance strategies. MGI's analysis of the impact of social technologies across four manufacturing sectors—consumer packaged goods, semiconductors, automotive, and aerospace—shows potential margin improvements of 2–6.5 percentage points, providing companies can transform traditional manufacturing IT into an all-encompassing information strategy to fine-tune product requirements, improve manufacturing processes, and boost quality and productivity.

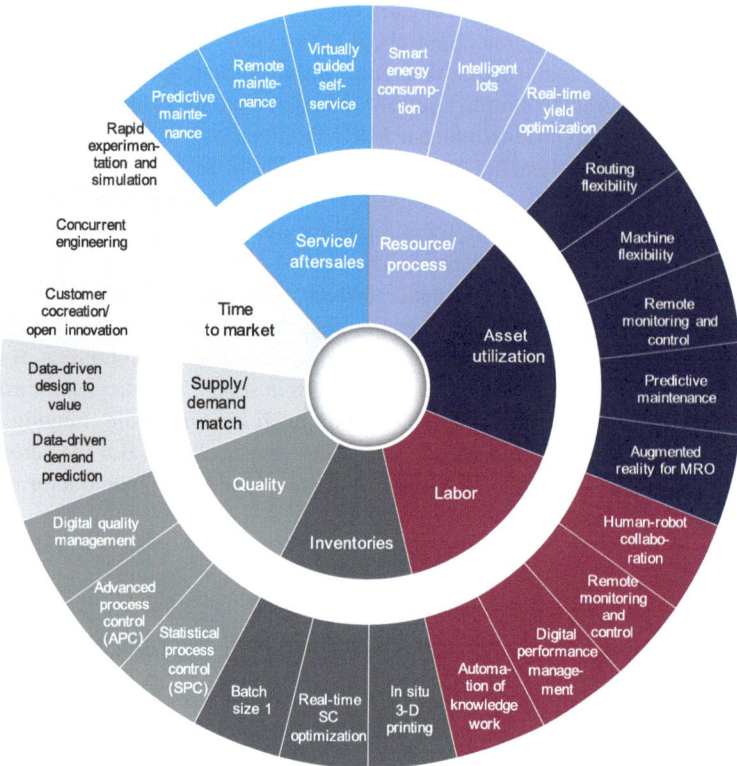

Fig. 3 A "digital compass" can help firms create and monitor the Digital Thread in their end-to-end operations. Source: Cornelius Baur and Dominik Wee, "Manufacturing's next act", McKinsey Quarterly, June 2015

These trends are already becoming evident in the commercial aerospace industry. Leading OEMs have invested in digital tools to integrate and manage their enormously complex supply networks. In the long aerospace supply chain, a Digital Thread allows companies to gain greater visibility into the end-to-end supply network and to track orders and changes throughout the chain. This is of significant value, since the complexity of sourcing can multiply quickly as one design modification impacts the manufacturing of many other components. With greater digital visibility, companies may find it easier to virtually integrate and monitor the supply chain, potentially accelerating a shift to low-cost sourcing. The industry already has examples of aerospace firms operating customers' fleet equipment for them; with more digitization and advanced analytics, companies may be able to operate more diverse products as well, creating new business models and revenue streams. Digitization can also enable aerospace suppliers to collaborate faster and more efficiently and to co-innovate with OEMs. For instance, an engine manufacturer can share detailed three-dimensional models with suppliers; each supplier can then

Trends in the Commercial Aerospace Industry 155

share information about price, delivery, and quality. This type of information sharing and transparency reduces the labor required to manage design changes and reduces risk for the engine maker and suppliers.

6 Policy

Government support has been critical for the growth of the commercial aerospace industry over the decades. Today the aerospace industry continues to see more regulatory and policy support than many other manufacturing industries. Government support can be in the form of building enablers—research funding and facilities, infrastructure development, workforce training, and various forms of financial support. It can also be in the form of actions to tilt the playing field with technology transfers. With changes in demand patterns, the industry is seeing more policy action on offset clauses and demand for technology transfers, for example in emerging markets such as China and India, to promote domestic aerospace manufacturing sectors. Government support can also come in the form of fostering national champions, supporting operations, and intervening to set up or restructure the sector.

6.1 Governments Continue to Build Enablers for the Aerospace Industry

Governments continue to invest heavily to provide aerospace with a supportive economic environment. One way is by building infrastructure; the state of Alabama's support for Airbus, for instance, includes road construction to support a new facility. Singapore also invested heavily in developing an integrated aerospace industrial park at Selatar Airport in order to attract international aerospace manufacturers. Infrastructure support can also include workforce training, and often these initiatives go back decades. In 1950, for instance, Brazil's establishment of the Instituto Tecnológico de Aeronáutica helped Embraer overcome a lack of core aerospace design capabilities. Today, many governments, especially in developed economies, focus on STEM (science, technology, engineering, math) training as a core element of support for the industrial sector.

Another 'enabler' approach is to support innovation, fostering or funding R&D activities to promote technological advances or improve access to new technologies. Over the past two decades, Canada has established research facilities such as the National Research Council Institute for Aerospace Research and the Strategic Aerospace and Defense Initiative, and the government continues to encourage collaboration between universities and industry to support the domestic aerospace industry. The UK government established a series of manufacturing research centers in collaboration with industry partners, and in 2012, it announced an £80 million package to "ensure the UK stays at the forefront of advanced manufacturing ... particularly in aerospace."(UK government, Department for Business, Innovation and Skills)· In Mexico, the state government of Querétaro reinforced its attractiveness as an aerospace cluster with a range of enablers—investing in an industrial park, collaborating with the government of Quebec to set up a training program, and pushing for a bilateral aviation safety agreement between Mexico and the United States to enable local certification of Mexico-produced aerospace goods.

6.2 Policy Initiatives Are Tilting the Playing Field

Government policy to support domestic aerospace manufacturers has also been, and continues to be, in the form of actions to tilt the playing field by creating incentives for domestic production or stimulating local demand. Some of these actions amount to direct financial support: Airbus, for instance, received more than $150 million from Alabama in financial and logistical support, including more than $60 million for equipment and capital costs and more than $50 million for workforce training.

With changes in demand patterns, the industry is seeing more policy action riding on market access and demand growth. China's ARJ21 and C919 programs were both launched on the back of orders from domestic airlines, many of which are government-controlled. Across the emerging world, many national airline companies remain wholly or partly under government ownership, allowing governments another way to use domestic demand to tilt the playing field. Commercial offset clauses and demand for technology transfers, particularly in emerging markets such as China and India, have been used to promote the domestic aerospace manufacturing sectors. When Airbus announced an extension of the assembly line contract with its Chinese partners, the company also announced an order for 70 aircraft from the China Aviation Supplies holding company (Flotau Jens 2014) Airbus extends Tianjin final assembly line contract.

Historically, many governments have been actively involved in setting up and expanding national aerospace champions, and the trend continues today. Embraer was founded in 1970 as a government-owned company. The Italian government continues to own about 30% of Leonardo. The Chinese government led the establishment of COMAC as a way to promote the domestic commercial aerospace industry. This approach can also take the form of industry restructuring. Airbus, for instance, was formed in 1969 as a consortium of Sud Aviation (France) and Deutsche Airbus (Germany), joined in later years by CASA (Spain) and Hawker Siddeley (UK). Today the 'national champion' approach continues to shape the corporate landscape of the aerospace sector. Some aerospace firms, such as India's HAL, are directly government-owned. Others are family-controlled but are considered national assets, including Korea Aerospace Industries, Japan's 'heavies' (such as Fuji Heavy Industries and Mitsubishi Heavy Industries), and Canada's Bombardier, recipient of $1 billion of support from the Quebec government.

7 Conclusion

Trends in the commercial aerospace sector point to continued expansion of the footprint into emerging markets on the back of 'super cycle'-driven growth. Demand growth is shifting to emerging markets, and the Asia-Pacific region could soon become the world's largest aviation market. Supply chains are already established in the region; Western incumbents have set up capacity, embarked on low-cost sourcing, and sought MRO partners in the local aftermarket. The rise of local OEMs will further accelerate the deepening of local supply chains, notably in China, which has doubled its share of global value added to 8 % in just the past 5 years. Other emerging markets, particularly smaller ones such as the 'Next 11' countries, are less likely to see the rise of a homegrown commercial OEM but could see more investment in the supply chain. Some of these countries, such as Korea, are already home to domestic aerospace manufacturing firms. As their aviation markets expand, MRO activities ramp up, and investment and expertise grow. Other countries—especially those with an established manufacturing or supply chain base, such as Mexico, Indonesia, and Vietnam—could see more sourcing of components and, in some cases, subsystems as aerospace firms look for transparency, risk-sharing, and partnerships with their suppliers.

Given the long time investment (5–7 years) to build up capacity and the risks that incumbents face in intellectual property, quality, and certification, these shifts in the aerospace industry may continue to lag the footprint shifts of other manufacturing industries such as automotive. North America and Western Europe make up over 80 % of value added in aerospace, while they account for less than 50 % of value added in the manufacturing sector overall. But the shift of the global aerospace footprint to Asia-Pacific could speed up in some scenarios. For instance, if newer competitors are able to put pressure on margins and gain market share, that may accelerate incumbents' search for lower-cost inputs and suppliers as they seek to preserve their profit margins—particularly in fragmented parts of the value chain such as aerostructures, where margins tend to be lower than for OEMs or engine manufacturers. This trend is often seen in a saturated market; it is less evident in markets in a strong demand cycle such as commercial aerospace, but it could still occur. Another factor that could accelerate the shift is digital technology. The deployment of a Digital Thread throughout the aerospace value chain may allow OEMs and Super tier-1s to better track their supply chains and production locations, focus on design and systems integration activity, and develop new business models in equipment operations. That could mitigate some of the risks to local sourcing and drive incumbents to outsource more physical activities, such as component production and distribution, while building up capability and value creation in design, analytics, and integration.

References

Airbus press release, 26 Feb 2014
Airbus (2016) Airbus Global Market Forecast, 2015–2034. Available via http://www.airbus.com/company/market/forecast/. Accessed 22 Jun 2016
Aviation Week (2015) Supply chain research insights: global aerospace industry size and growth. Available via http://aviationweek.com/master-supply-chain/supply-chain-research-insights-global-aerospace-industry-size-and-growth. Accessed 22 Jun 2016
Baur C, Wee D (2015) Manufacturing's next act. McKinsey Quarterly, Jun 2015. Available via http://www.mckinsey.com/business-functions/operations/our-insights/manufacturings-next-act. Accessed 22 Jun 2016
Boeing press release, 23 Sep 2015
Boeing (2016) Boeing current market outlook, 2015–2034. Available via http://www.boeing.com/resources/boeingdotcom/commercial/about-our-market/assets/downloads/Boeing_Current_Market_Outlook_2015.pdf. Accessed 22 Jun 2016
Cohen D, George K, Shaw C (2015) Are you ready for 3-D printing? McKinsey Quarterly, Feb 2015. Available via http://www.mckinsey.com/business-functions/operations/our-insights/are-you-ready-for-3-d-printing. Accessed 22 Jun 2016
Flotau Jens (2014) Airbus extends Tianjin final assembly line contract. Aviation Week Network. Available via http://aviationweek.com/awin/airbus-extends-tianjin-final-assembly-line-contract. Accessed 22 Jun 2016
George K, Ramaswamy S, Rassey L (2013) Next-shoring: a CEO's guide. McKinsey Quarterly, Jan 2013. Available via http://www.mckinsey.com/business-functions/operations/our-insights/next-shoring-a-ceos-guide. Accessed 22 Jun 2016
Global Wind Energy Council (2015) Global wind energy outlook. Available via http://www.gwec.net/wp-content/uploads/2014/10/GWEO2014_WEB.pdf. Accessed 22 Jun 2016
Hartmann B, King WP, Narayanan S (2015) Digital manufacturing: the revolution will be virtualized. McKinsey Quarterly, Aug 2015. Available via http://www.mckinsey.com/insights/operations/digital_manufacturing_the_revolution_will_be_virtualized. Accessed 22 Jun 2016
Manpower Group Talent Shortage Survey 2015
McKinsey & Company (2012) Winning the $30 trillion decathlon: going for gold in emerging markets. McKinsey Quarterly, Aug 2012. Available via http://www.mckinsey.com/business-functions/strategy-and-corporate-finance/our-insights/winning-the-30-trillion-decathlon-going-for-gold-in-emerging-markets. Accessed 22 Jun 2016
McKinsey Global Institute (2013) Urban World: the shifting global business landscape. McKinsey Global Institute Report, Oct 2013. Available via http://www.mckinsey.com/global-themes/urbanization/urban-world-the-shifting-global-business-landscape. Accessed 22 Jun 2016
McKinsey Global Institute (2015) Playing to win: the new global competition for corporate profits. McKinsey Global Institute Report, Sept 2015. Available via http://www.mckinsey.com/business-functions/strategy-and-corporate-finance/our-insights/the-new-global-competition-for-corporate-profits. Accessed 22 Jun 2016
McKinsey Global Institute and McKinsey Operations Practice (2012) Manufacturing the future: the next era of global growth and innovation. McKinsey Quarterly, Nov 2012. Available via http://www.mckinsey.com/business-functions/operations/our-insights/the-future-of-manufacturing. Accessed 22 Jun 2016
UK Government's Department for Business, Innovation & Skills (2012) Lift off for aerospace and manufacturing projects. Press release, Jun 2012. Available via https://www.gov.uk/government/news/lift-off-for-aerospace-and-manufacturing-projects. Accessed 22 Jun 2016

Gernot Strube is a director (senior partner) in the Munich office of McKinsey & Company. Since joining the firm in 1993, Gernot has worked mainly with industrial corporations in aerospace, machinery, high tech, and automotive sectors. Over time he has been based in McKinsey's offices in Cleveland, Hong Kong, and Munich, and has served clients in over 25 countries. While his work includes a broad range of topics, he focuses on operational strategy and performance improvement programs. Gernot earned a M.Sc. in aeronautical engineering and holds a Ph.D. in mechanical engineering from the University of Technology, Munich, Germany. Prior to joining McKinsey, he worked for five years as a scientific researcher in the field of applied laser diagnostics in combustion engineering for governments and leading automotive and aerospace companies.

Karel Eloot is a director (senior partner) in the Shanghai office of McKinsey & Company. He joined the firm in 1997 in the Benelux office before transferring in 2006 to China. He is a co-leader of McKinsey's operations practice in Asia-Pacific. He serves clients in automotive, aerospace, high tech and basic materials sectors. His work includes manufacturing and operations transformation, product development, supply chain management, sourcing, commercial and organizational work, strategy development, and mergers and acquisitions. Prior to joining McKinsey, Karel worked as a steel researcher in Belgium, Germany, the United States, and Japan, authoring papers for international journals and receiving awards for his work. He received the degrees of metallurgical engineer and Ph.D. in materials science from the University of Gent, Belgium.

Nadine Griessmann is the global knowledge manager for McKinsey & Company's aerospace and defense practice, and is based in Munich. Since joining the firm in 2007 as an industry analyst, she has worked with clients in the aerospace sector on corporate and business unit strategy, international growth, and new business models. She has co-authored and contributed to several of McKinsey's aerospace and defense-related publications, including *The future of European defence: Tackling the productivity challenge* (2011), *Growth and renewal in the United States: Retooling America's economic engine* (2011), and *Southeast Asia: The next growth opportunity in defense* (2014). Prior to joining McKinsey, Nadine worked as an analyst in the aerospace industry. She has Master's and undergraduate degrees in international business administration.

Rajat Dhawan is a director (senior partner) in the New Delhi office of McKinsey & Company. During the past two decades at the firm, he has worked in several regions across Asia, Europe, and North America. He serves clients primarily in automotive, aerospace and defense, and other advanced industries, and leads McKinsey's operations practice in the Asia-Pacific region. Rajat focuses on themes of strategy, business-building and operational excellence. He advises business and policy leaders in Indian manufacturing; he is the founder director of the McKinsey Capability Center, and former member of the Manufacturing Council of the Confederation of Indian Industry. Over the years he has partnered with several automotive industry associations to help catalyze industry growth. Rajat has an M.B.A. and B. Tech. degree in chemical engineering.

Sree Ramaswamy is a senior fellow with the McKinsey Global Institute (MGI), McKinsey & Company's business and economics research arm. Since joining McKinsey in 2008 in Washington DC, he has served clients in industrials, high tech, and energy sectors. He leads MGI's research on the role of multinational firms in the global economy and the impact of competition and technology on economic performance. He co-leads MGI research on manufacturing and supply chains in partnership with McKinsey's operations practice, and is a frequent speaker on MGI research topics. Sree earned an M.B.A. from Columbia University and Master's and undergraduate degrees in telecommunications and computer engineering. Prior to joining McKinsey, he worked for ten years in the U.S. aerospace industry and holds three patents in satellite systems design.

Success Through Customer Co-Development, Global Footprint and the Processes In-Line with the Customer

Mark C. Hiller and Joachim Ley

Abstract At a first glance, comparing automotive and aviation industry is difficult since, for example, the number of car seats produced per day matches the number of aircraft seats produced per year. However, by investing some efforts to further analyze commonalities, many areas can be found where the industries can benefit from each other.

When it comes to operational processes and global footprints, the automotive industry is, and will most probably remain ahead, thus being a role model for the aviation industry. On the other hand, aviation always had a strong focus on light weight design which has gained more importance in the automotive industry in recent years.

The case study in this article describes how RECARO Aircraft Seating combined their proprietary knowledge from the aviation industry with knowledge transfer from the automotive industry to become a leading aviation supplier and a reliable partner to their airline customers and the OEMs.

In order to achieve this, RECARO has identified six success factors: product development philosophy and innovation, product architecture, project management, global supply chain, a holistic "lean" approach as well as an extended enterprise to cope with future challenges. The contribution of each of these factors is introduced in the case study.

1 RECARO Aircraft Seating and the Automotive Heritage

1.1 The Heritage of RECARO

It all started in 1906 when Wilhelm Reutter established "Stuttgarter Carosserie-u. Radfabrik" (Stuttgart Body and Wheel Factory). The company quickly became

---

M.C. Hiller • J. Ley
RECARO Aircraft Seating Gmbh & Co.KG, Schwäbisch Hall, Germany
e-mail: Mark.Hiller@recaro-as.com; Joachim.Ley@recaro-as.com

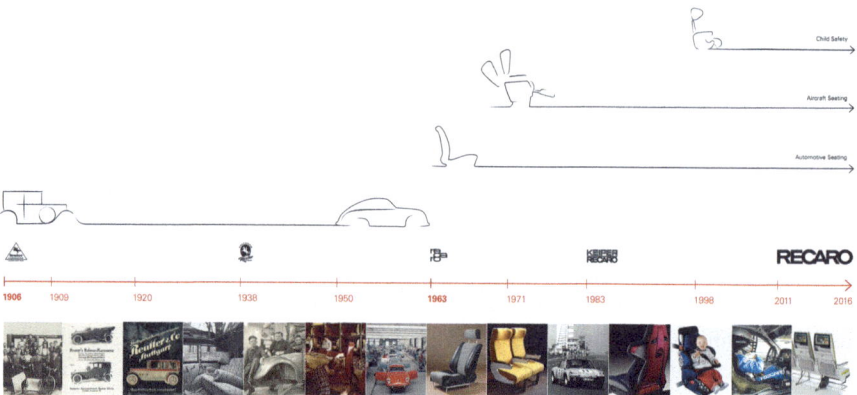

Fig. 1 Milestone of the RECARO Group. Source: RECARO Group

an important force in the young automotive industry. Reutter enjoys success with the "Reform-Karosserie", a predecessor of today's convertible, which was already patented in 1909. The company successively delivered individual and serial bodies for nearly all renowned car makers at that time, always with the complete interior fittings, including seats. Till 1963 the company produced the majority of the Porsche 356 bodies (RECARO 2015).

In 1963, the company re-invented itself. The car body factory was sold to Porsche but the seating expertise remained within the company. The combination of REutter and CAROsserie lead to the name RECARO (2015). The DNA of the RECARO products is a combination of functionality, aesthetics and ergonomics which RECARO calls "ingenious design". RECARO is one of the best known brand names, if not the best known name for mobile seating products. In 1971, the company started stepping into the aviation industry and to apply ingenious design to aircraft seats. In 1998, the RECARO portfolio was expanded by the child seat business (Fig. 1).

1.2 Learning from the Automotive Industry for the Aviation Industry and Vice Versa

It is possible to derive lessons from one to the other industry and to realize synergies regardless of fundamental differences between them. Lessons can be learned from the automotive industry especially for aircraft seating in terms of global footprint and industrialization.

Let's compare the business volumes between automotive and aviation for seats: the automotive business, was about 92.53 million vehicles worldwide in 2015 (Statista 2016). If we moderately calculate just two seats per car, the daily output of seats for 100 million vehicles would be around 550,000 seats per calendar day. The production output from Airbus and Boeing in 2015, in comparison, was around 1400 aircraft (Wikipedia). Assuming even 250 seats per aircraft on average, this would lead to a demand of 350,000 seats per year. Considering an additional 60% demand increase for existing aircraft fleet seat retro-fits, this would amount to 560,000 seats per year. We can conclude that the daily car seat production equals the yearly production of aircraft seats. Due to the volume, the automotive industry has a global presence and is hence the role model.

Besides footprint and industrialization, there are other areas for synergies and lessons to learn. In the automotive industry, simulations for seat crash behavior in terms of reliability and manufacturability are very much advanced. The aircraft seat certification, on the other hand, is still heavily relying on real tests. Simulation is mainly used for better predictability of the real tests. Therefore, certification based on simulation is a future potential for tremendous reductions in lead-times and in aircraft seat development cost.

In sophisticated car rear seat projects, for example, automotive seating is learning from aviation seating. The comfort of an aircraft business class seat from RECARO was the blueprint for the rear seating concept of a luxury car. When seats should support passengers not just in seating positions but also in relaxing, dinning or sleeping positions, ideas and concepts can be transferred from aviation to automotive. The progress towards autonomous driving will further shift the focus from providing pure seating in automotive towards concepts which are closer to an aircraft business class environment. There are also other areas in the more short term where aviation can provide useful guidance to automotive. Aircraft seats are designed for lowest weight and space consumption. With the focus on efficiency, light weight construction becomes more and more important, also for electrical cars, where heavy batteries need to be compensated by other lighter components. Aircraft seats have become more advanced in terms of space efficiency and slim design in the last years. This could benefit automotive with either better comfort through more living space, or the same comfort with smaller dimensions providing potential for weight and cost reductions.

2 What Are the Chances and Challenges of an Aircraft Seating Supplier?

2.1 Strong Market Growth Worldwide

Derived from Airbus' and Boeing's market forecast (Airbus 2015; Boeing 2015a), the graph shows a market forecast for aircraft seats in Mio. € by region. The volume includes different kinds of seats for different classes (from economy to first class) and for line-fit as well as retro-fit. Seating retro-fitting normally takes place after 7 years of operation (Fig. 2).

MARKET SIZE AND GROWTH FOR AIRCRAFT SEATS FROM 2015 TO 2020 BY REGION

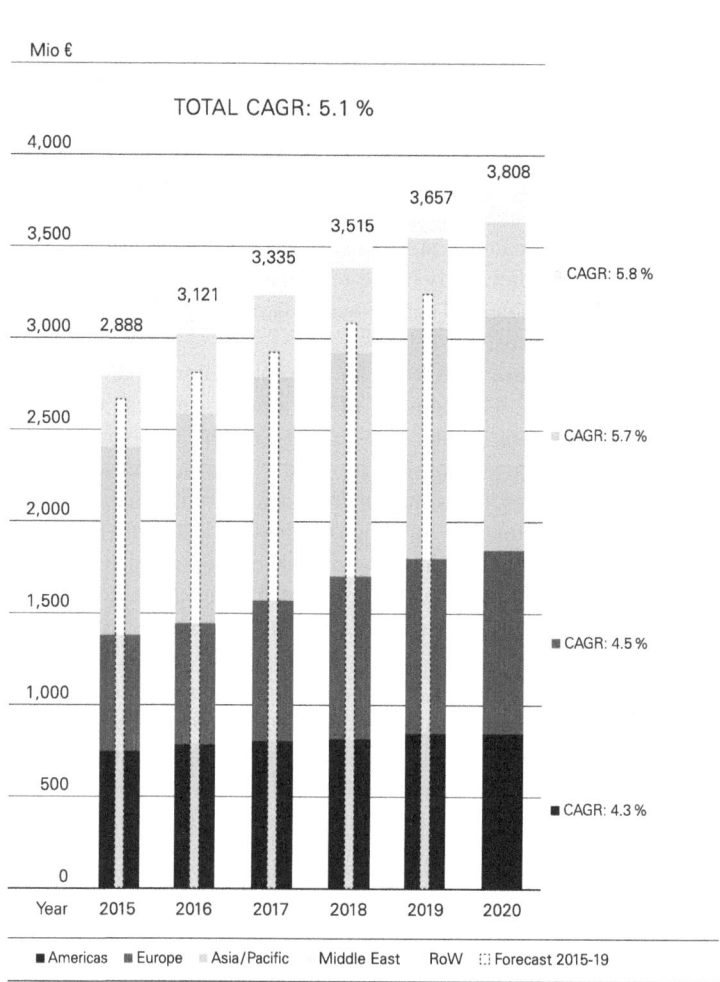

Fig. 2 Market size and growth for aircraft seats from 2015 to 2020 by region

For a component supplier in the aviation industry, the diversity of the world market in terms of *market regions*, *airlines business models* and *aircraft types* is a great opportunity.

First of all, the *market regions* have different maturity levels. In North America, where privatization and consolidation took already place, there are just a few, yet major airlines left with the largest fleets in the world. In Europe, most airlines are privatized, but the consolidation is still in progress and might not go as far as it went in North America. In Asia, and especially in China, we have a lot of state owned airlines and new airlines are still being set up.

In different market regions, airlines are also operating different *business models*. There are legacy carriers as well as LCCs (low cost carriers). Today, there are also mixed models. Legacy carriers operate selected business units on certain routes as low cost models. LCCs, already a significant portion of the market, rely on ancillary revenue (revenue beyond the pure sales price of a ticket such as baggage fees, on-board food and services) which has become a relevant financial component. It can count for more than 20 % of the airlines' revenue (Sorensen 2016). For a component supplier, the challenge is not only to supply an innovative, reliable and reasonable priced component. It is even more important to support the airline in generating additional customer value which contributes to ancillary revenue.

The strong market growth in aviation is not just supported by an increased output of existing aircraft types at the established OEMs. The numbers of available *aircraft types* will increase significantly with new, much more fuel efficient aircraft from existing OEMs as well as new aircraft from new OEMs (Boeing 2015b). These will on the one hand support the market demand, on the other hand, they will stimulate additional demand. The hurdle for replacing existing aircraft in the fleet is much lower if newer and more efficient aircraft are available.

Aircraft interior plays a very important role for the airline. It is one of the main differentiators in competition against other airlines, besides routes, prices and services (SKIFT Report #23 2014). In contradiction, the interior, especially the aircraft seats, account for less than 5 % of the purchase price of the whole aircraft. Yet the airline draws a lot of attention to the seats. They can be replaced easily compared to other aircraft parts over the aircraft lifetime in order to renew the customer experience, or to increase the efficiency.

2.2 Variety of Customer Types

Most interior suppliers have more than one customer type, such as OEMs purchasing SFE (supplier furnished equipment) and airlines purchasing BFE (buyer furnished equipment) for their different airline types (LCC or legacy), representing the majority of the seating business. There are also leasing companies who account nowadays for almost 30 % of the orders for Airbus and Boeing (Airbus 2015). Because of the coexistence of the different customers, an interior supplier is not able to choose between them, he needs to satisfy all requests from the different customers. Nevertheless, this can drive very different and even contradictory requirements.

OEMs, for example, especially in a time with a major focus on the ramp up, have a strong focus on a short aircraft interior implementation time. Otherwise the cycle times in the final assembly lines cannot be reduced. Late changes are also not of interest to an OEM because of the risk for the supply chain.

For airlines, the assembly process is not a special focus. If there are innovations which create a major differentiation in the tough competition, airlines are willing to implement even late changes. The aircraft delivery is more seen as a single event, while airlines generally buy aircraft with the intention of operating them for up to 30 years, the interior will be in service at least until the first retro-fit cycle starts after 7 years. Communality is another factor with great importance to an airline. New aircraft are becoming part of an existing fleet. To support interchangeability of new aircraft with the existing aircraft, cabin communality, such as same seats, is needed. Innovations need to be available across aircraft types and existing seats need to be upgradable, if not replaced, with features of the new generation. When airlines started to integrate IFE (in-flight entertainment) into new aircraft seats, for example, there was a strong demand for also upgrading the existing fleet seats with IFE.

Leasing companies have short turnaround times, hence the time between receiving an aircraft from the lessee until handing it over to the next lessee, needs to be minimized. Concepts for maintaining, or even upgrading the interior within short time frames are important.

2.3 Variety of Products

In addition to the variation between customer types there is also a big seat variety within one seat type for one aircraft layout, called LOPA (layout of passenger accommodation). The aircraft layout can be seen as one of the most expensive real estates. The optimal space utilization has highest priority over standardization. A LOPA with 350 economy class seat places can consist of around 100 seat benches, including double, triple and quadruple seats. These 100 seat benches have 50 different part numbers because there are special seats such as exit seats, first row seats, last row seats, as well as left hand, center and right hand positing. The seat bench itself consists of a couple of hundred parts. The combination of small volume in aviation compared to automotive (see Sect. 1.2) and a high number of parts requires a sophisticated variant management and product architecture.

3 A Global Footprint Tailored to OEMs and Airline Customers

Analogies between automotive and aviation can be found in regards to global footprints and global supply chains. This chapter briefly describes the automotive principle and how RECARO applies this to their global network.

Decades ago, the OEMs in the automotive industry decided to follow their customers. High volume car manufacturers moved their factories and supply chains close to their markets. There are only few exceptions such as smaller car manufacturers producing luxury or niche models. Once it is decided that a plant will be set up at a certain location, at least some of the suppliers will follow. When it comes to car seats, there is always a seating plant in close proximity to the OEM since they are usually supplied just in sequence straight to the production line. This requires very mature and robust processes because the seat manufacturer is often only informed several hours prior to the car seat installation about the configuration and the sequence of the seats (Liker 2009).

Major markets in civil aviation are Asia/Pacific, North America and Europe. In total they will generate 85 % of the global demand of new airplanes up to 2034 according to the latest Airbus global market forecast (Airbus 2015). Until the end of the last decade, both major OEMs built airplanes in one, respectively two locations in North America and Europe. In the meantime they also decided to follow their customers. Airbus has established final assembly lines in Europe, China and North America and Boeing will follow as well with a completion and delivery center in China as announced in September 2015 (Boeing 2015c) (Fig. 3).

RECARO's footprint perfectly matches the market as well as the OEM's footprints with assembly plants in Germany, Poland, the USA and China. RECARO is headquartered in Germany: the site in the US was set-up in the late 1990s, the Poland plant in 2006, and the China plant in 2013, each of them with a distinctive profile.

RECARO's headquarters in Germany has all engineering, supply chain, operations and after sales service capabilities. A highly qualified and experienced cross functional team develops and manufactures all new products, from the initial sketch to first customer programs in order to ensure a smooth ramp-up. Having all disciplines at the site in close proximity, the products quickly reach a high maturity level and can then also be transferred to other sites.

The first international plant was established in the US, at that time driven by a major customer order from a North American airline. It was set-up next to the customer's maintenance center which generated benefits in logistics and lead-time

Fig. 3 Overview on major market regions, OEM and RECARO footprint

with minimized foreign exchange rate risks. Initially the plant was connected to the existing supply chain located mainly in Europe. Today, this plant is important for supplies to the American retro-fit market and to the local OEMs. The supply chain has been localized to a large extend, further eliminating foreign exchange risks through natural hedging. The plant is capable of handling customization programs locally since it is a fully fledged site including customization engineering.

Compared to the US, the Poland approach was different. Competing on a global market, there is a lot of cost pressure, especially in the high volume segments. The plant is run as an extended workbench to the German facility with minimum overhead structures. It only produces seats for one segment. Combined with a holistic "lean" approach, this enables a highly efficient plant with very mature processes and stable high quality products. Employing the existing supply chain in the beginning, more and more parts are purchased locally in the meantime. Together with the German plant it serves the European retro-fit market as well as the European OEMs on a reasonable cost base.

The Chinese plant was set up with a purpose similar to the US approach. Targeting the market in Asia/Pacific, the plant is equipped with engineering capabilities for customization. This is important to customers since there is typically a lot of communication during the project which can take place without the burden of huge time shifts and language barriers. Furthermore, all lessons learned from the other plants were incorporated in the set-up process. This especially applies to the supply chain. The full benefit in terms of purchase price, transportation cost and lead-time relies on a local supplier base to a huge extend. In the past, the plant was always set up first and the supply chain followed. In this case it was done vice versa. More than 2 years prior to the plant set-up, a purchasing office was established in order to identify and qualify local suppliers. This enabled the plant to start with more than 60 % local purchases, having the full benefits right from the start.

All plants have a distinctive profile concerning market approach and product portfolio. Products with low volumes and high complexity, for example, which require more support during their lifecycle are manufactured at the sites with sufficient overhead available. Less complex high volume products are manufactured in the Poland plant which offers the highest efficiency. All core processes are globally harmonized enabling RECARO to shift programs from site to site on short notice. Process owners are spread across all locations. This ensures always best process employment, which is not necessarily carried out by a centralized team.

The global network has several advantages. It enables RECARO to serve the customers in their time zone, to offer attractive lead-times through proximity to the OEMs and airline customers, to use natural instead of financial hedging, to mitigate the risk for the customer since programs can be shifted from one site to another site, and finally to deliver consistent quality through globally standardized processes.

The global footprint is the foundation for setting up successful business in aviation. However, it also requires the right processes and methods to exploit the full network potential. RECARO has identified six so-called enablers for their company.

3.1 Product Development Philosophy and Innovation

Up until the beginning of this decade, the product development process at RECARO had been mainly driven by single customers. If a customer was interested in a new seat, RECARO worked intensively together with them, with very major efforts to achieve the defined targets. High risks were taken, because innovations in terms of new technologies, new materials or completely new features were integrated in customer projects. The developed products met the requirements but were very often limited to just single customer requests. Other customers' requests could not be met because of the tremendous time pressure within the one customer project. To be able to serve other customers, major modifications or even complete new developments were needed. A paradigm shift was necessary in order to achieve a higher innovation rate by reducing the customer program risk and by increasing the efficiency at the same time. In 2010, RECARO decided to shift from developing products for single customer requests to modular product development for market needs. In consequence, the entire product development process (see Fig. 4) and the development organization were restructured.

With the new market oriented philosophy there is a strong emphasis on front loading with idea creation based on scenarios, trend analysis, idea scouting and QFD (quality function deployment) workshops with potential customers. Afterward the ideas are evaluated and selected. Selected ideas are becoming part of the technology roadmap, which describes the duty book of the innovations. If the idea is successfully developed to a certain maturity level, it will be available in the innovation shelf for the further development process.

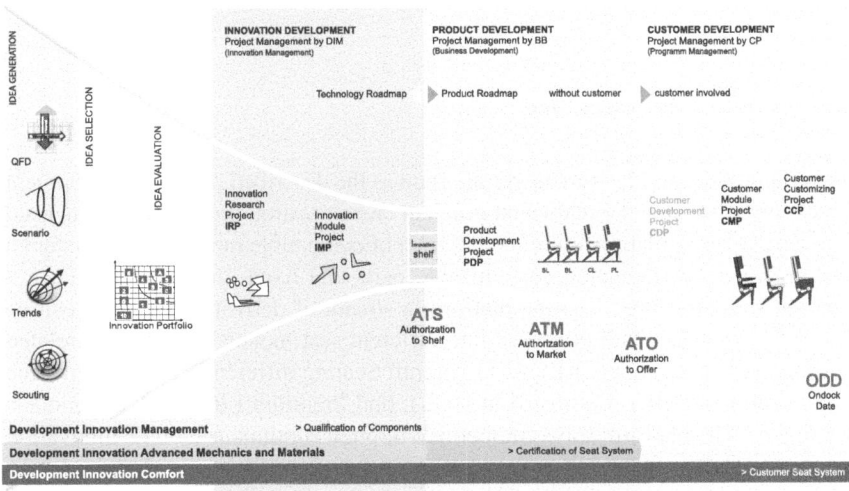

Fig. 4 Product development process at RECARO—from initial ideas to customer development projects

In the platform development process, new products can be developed based on specifications without a customer order. Customers might be involved as sparring partners: to receive detailed feedback on new concepts and technologies and to avoid developments which will not meet the customer demands. The platform development process ends when the product platform maturity is achieved, reaching specific deliverables as well as successfully passing predefined test procedures. The milestone at the end of the process, named ATO (authorization to offer), is also the starting point for the acquisition of first customer orders and the customer development process.

The results of this paradigm shift are: lead-times in customizing projects are shortened by almost 30 %, new product sales are increased from 15 to 30 % of total sales (products younger than 3 years) combined with a much lower introduction risk. The necessary customization effort per program was also significantly reduced, however, the precondition for that is the upfront investment which increased the R&D rate (spending for R&D in relation to total sales) from 6 % to above 10 %. Upfront investment also allows to better integrate suppliers into the value creation process right from the beginning. Ideas and innovations from suppliers can be considered early in the process and feasibility can be analyzed before concepts and drawings are finalized.

The implementation of the described development processes was achieved by a 3 years ongoing change management project and the introduction of a new development organization. The whole R&D organization with 350 employees, representing more than a third of the total German headquarters staff, was divided into three units: innovation unit with 60 employees, platform development unit with 180 employees and validation and verification unit with 110 employees. The latter is a service unit which provides simulation, testing, quality assurance and certification along the entire development process.

3.2 Product Architecture

The right product architecture is as important as the described product development process and structure. To realize an efficient customization process, products need to be developed as platforms, defined as a set of compatible modules (sub-systems) with specified and regulated customizable parts and fixed interfaces that form a common structure. Based on the platform a stream of derivative products can be efficiently developed and produced into different seat models to address a related set of market applications. RECARO Aircraft Seating differentiates in Smart Line (SL), Basic Line (BL), Comfort Line (CL), und Premium Line (PL) seat models (see Fig. 5). The platform stays competitive through continuous module innovation and renewal. A module consists of various related components. Module design changes should not impact other modules and must not affect the interfaces. It is desirable to carry over the same modules to different platforms (RECARO Aircraft Seating 2012). (Fig. 5)

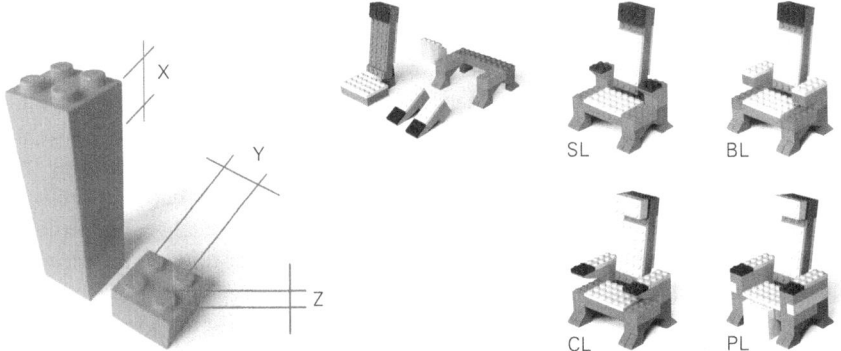

Fig. 5 Definition of interfaces and platforms

RECARO is combining and balancing both centralized and decentralized approaches, protecting product platform interfaces and the system intelligence. The platform development is exclusively located at the headquarters but all sites are entitled to develop modules and components to fulfill the needs of the locally run customization projects. Depending on the market demand, these modules and components can become part of the global product platform.

The modular product platform development leads to two major organization requirements: A holistic system understanding of the product platform and a very detailed knowledge of module and component development. The described unit platform development (see Sect. 3.1) is therefore divided into two departments: platform system and modules. The platform system teams are mainly coordinating the different product platforms and assure that the customization projects are within the platform boundaries. In the module organization are the specialists located which have very detailed knowledge about the different modules and components, such as headrests, backrests and armrests. The module organization provides also suitable interfaces to cooperate with purchasing and relevant suppliers.

3.3 Project Management

Even with the early application of project management to the space and aviation industry (Hiller 2003), project management is still seen as one very essential methodology to manage the contradictory requirements of the industry. On one hand, the market is asking for more and more differentiating products and solutions, on the other hand increasing demand requires shorter lead-times, highest delivery performance and quality. RECARO Aircraft seating implemented a worldwide standardized project management at the beginning of 2004. Since then, there has always been a dedicated project manager for every customer and platform development project. These projects are run by an eight phase model, having a detailed audit at the end of each phase. The audit is performed by employees which are not part of the project team in order to get an unprejudiced evaluation of the project status. To apply lessons learned from one project to the next one, a detailed questionnaire is used for the audit. There are specific questionnaires for the relevant audit phases. The PMO (project management office), a project management supporting team, takes lessons learned from every project to evolve the questionnaire by adding or modifying the questions. The project manager is presenting the audits' results to the steering meeting. Based on these results, a go or no-go decision will be taken. The steering meeting is taking place once a week. Different projects are presented once a phase ends, or if the project manager or the steering meeting request a status report.

In customizing projects, a status presentation excerpt is also used to inform the customer about the project status. This transparency helps to enable customers to take decisions necessary to proceed with the project.

3.4 Global Supply Chain

It will be difficult nowadays to find a company that does not claim to have a global supply chain. In most cases this is cost driven, which obviously also applies to RECARO. Having a robust supply chain is of utmost importance, especially when most parts are designed in house but manufactured by external suppliers. Besides cost and robustness, a supply chain should adapt to currency fluctuations on the customer side and enable short lead-times in terms of "lean management".

RECARO has established a global purchasing team which is located at all production sites. They are in charge of the local purchasing activities for their plant, but also manage their plant suppliers in the region according to the global purchasing strategy.

Risk mitigation is typically based on a second source strategy, ideally having at least two suppliers in different regions, balancing currency fluctuations. RECARO is paying a lot of attention to supply chain management including sub-tiers. Poor supply chain management is still one of the major reasons why companies fail in the aviation business.

Another important aspect is local sourcing. There are many commodities in an aircraft seat where lower purchase prices are offset by transportation cost if they have to be shipped over long distances. In these cases global footprints are even required in the sub-tier supply chain. Another aspect of local sourcing is lead-time. As mentioned previously, this is important in a project driven environment with a strong "lean" background, and still one of the most important operational challenges for aviation supply chains (Staufen 2015).

3.5 Holistic "Lean" Approach

"Lean" is more than just tools and methods, it is rather a culture of how organizations approach things and continuously thrive for improvement by identifying and eliminating "waste". As most companies do, RECARO started with "lean" tools and methods in operations. It dates back to 2007, the headquarters being the pilot.

This quickly led to impressive results which encouraged the management to take the next step and tackle the cultural aspect which makes "lean" sustainable. It becomes visible and tangible in the so-called "shop floor management", which is based on transparency through KPIs (key performance indicators) in the information centers. All managers use a board on which their teams can display and track the relevant processes based on KPIs. The teams meet in clearly defined cycles, from daily to weekly, in order to review the progress. Thus, deviations can be identified quickly and appropriate measures are taken immediately. During the implementation phase, experienced coaches trained the leadership team on how to become a leader in a "lean" environment, and the team received basic training, for example in problem solving (Fig. 6).

RECARO is transforming the company into a "lean enterprise". "Shop floor management" is not only implemented in operations, it has become a vital part of daily life in all functions and at all plants. Each plant has got a cascade of information centers to ensure that everybody is working according to the same principles, thus problems are identified and tackled at the appropriate level. Besides this, each and every process is continuously improved with value stream mapping, always trying to reduce lead-time, which is the result of robust and "lean" processes with minimum "waste" (Weiß et al. 2015).

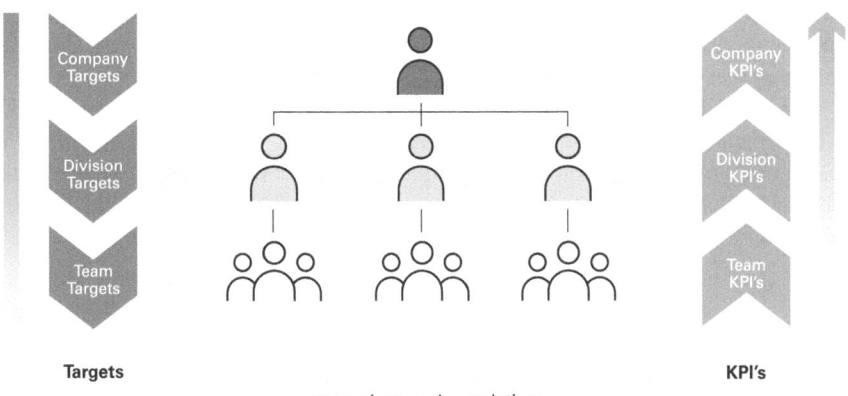

Fig. 6 Shopfloor management cascade

3.6 Extended Enterprise

For a smooth transition between the customer and the supplier value creation chain, it is not enough to just optimize within the boundaries of each company. The interfaces need to be defined and harmonized in the best possible way. One example where another concept from the automotive industry was taken to adapt and implement is the direct delivery process. Before the implementation of this direct delivery process between RECARO and Airbus for the A320 family production, shipsets (that are all seats for one aircraft) had been delivered according to the following steps:

1. pre-delivery seat inspection by Airbus on-site at RECARO
2. delivery in non-returnable packaging
3. incoming inspection at Airbus
4. storage of the seats in a central warehouse
5. delivery from warehouse to the final assembly line
6. sequencing of the seats
7. installation

After supply chain optimization, the steps are:

1. delivery of the seats in returnable packaging (this is used all the way, also for internal transportation at Airbus) already sequenced according to the installation order straight to the assembly line
2. short seat inspection with regards to potential transportation damages
3. seat installation

The optimization has reduced the lead-time by approximately 80 % with major reductions in processing time on both sides and less packaging material used. Preconditions for these achievements are the willingness of the customer and the supplier to optimize processes across company boundaries, and of course stable and reliable processes on both sides.

This is another example where RECARO benefits from its automotive heritage. In 1984, the predecessor company "KEIPER RECARO" started one of the first JIT (just in time) plants for automotive seating in Europe cooperating with Mercedes in Bremen (Fieten 1991). The direct delivery concept employed, it was again one of the partners' pioneering a similar concept in the aviation industry.

4 Conclusion

There are many ways of how aviation can benefit from automotive and vice versa. As the huge difference in terms of global production volume remains (see Sect. 1.2), it will most likely be the automotive industry staying ahead in terms of operational processes. If these processes are carefully analyzed and adapted, there are chances the aviation industry can obtain massive benefits by learning from the automotive industry for example, applying, global footprints with customer proximity or "lean" management (see Sect. 3.5). The RECARO Poland plant is a prototype for producing cost sensitive products with low complexity and high volumes (see Sect. 3.5).

On the product side it is expected to be vice versa. This is mainly driven by the increasing focus on weight reductions and shifting demands to an automotive seating when it comes to autonomous driving. Light weight design has always been important in aviation and is becoming more important in automotive, for example due to the weight of batteries in electric cars and rising fuel prices in conventional cars. It is expected that autonomous driving requires car seats similar to business class airplane seats for relaxing, working or even sleeping (see Sect. 1.2).

The automotive heritage as well as a strong focus on innovation, ingenious design and robust supply chains with customer proximity has helped RECARO to stay on top of the challenges of the aviation industry in the past years. It is a sound foundation for future success, but this needs to be continuously enhanced in order to defend this position.

References

Airbus (2015) General Market Forecast flying by numbers 2015–2034. Available via http://www.airbus.com/company/market/orders-deliveries/. Accessed 22 Jan 2016

Boeing (2015a) Current market outlook 2015–2034. Available via http://www.boeing.com/resources/boeingdotcom/commercial/about-our-market/assets/downloads/Boeing_Current_Market_Outlook_2015.pdf. Accessed 24 Jan 2016

Boeing (2015b) Homepage. Available via http://www.boeing.com/commercial/. Accessed 22 Jan 2016

Boeing (2015c) Boeing hosts China president Xi Jinping, announces airplane sales, expanded collaboration with China's aviation industry. PRNewswire, 23 Sept 2015. Boeing. Available via http://boeing.mediaroom.com/index.php?s=20295&item=129523. Accessed 24 Jan 2016

Fieten R (1991) Erfolgsstrategien für Zulieferer: Von der Abhängigkeit zur Partnerschaft Automobil- und Kommunikationsindustrie. Springer Gabler, Wiesbaden

Garcia M, Skift Team (2014) The future of the aircraft cabin: tracking trends and debunking myths. Available via http://skift.com/wp-content/uploads/2014/09/23-SkiftReport-The-Future-of-the-Aircraft-Cabin.pdf. Accessed 22 Jun 2016

Hiller M (2003) Multiprojektmanagement—Konzept zur Gestaltung, Regelung und Visualisierung einer Projektlandschaft. University of Kaiserslautern, Kaiserslautern

Liker J (2009) Der Toyota Weg: 14 Managementprinzipien des weltweit erfolgrichsten Automobilkonzerns, 6th edn. Finanzbuch Verlag, München

RECARO Aircraft Seating (2012) Engineering Excellence Project

RECARO (2015) The guiding principles of the RECARO Group. RECARO homepage. Available via http://en.recaro.com/recaro-world/tradition.html/. Accessed 20 Dec 2015

Sorensen J (2016) Ancillary revenue report series for 2015. The car trawler yearbook of ancillary revenue. Available via https://www.cartrawler.com/ct/assets/media/Downloads/PDFs/2015-Ancillary-Revenue-Yearbook.pdf. Accessed 10 May 2016

Statista (2016) http://de.statista.com/statistik/daten/

Staufen AG (2015) Aviation-Studie 2015: Globale Wertschöpfungsketten managen. Available via http://www.staufen.ag/fileadmin/hq/survey/STAUFEN-studie-aviation-globale-wertschoepfungsketten-managen-2015.pdf. Accessed 22 Jun 2016

Weiß E, Strubl C, Goschy W (2015) Lean Management: Grundlagen der Führung und Organisation lernender Unternehmen. Erich Schmidt Verlag, Berlin

Wikipedia. https://en.wihipedia.org/wiki/competition-between-Airbus-and-Boeing

Mark C. Hiller, Dr., studied Economic Science at the University in Kaiserslautern/Germany and at the Babcock Graduate School of Management in the USA. After various consulting projects at the Centre for Production Technology at the University of Kaiserslautern, he obtained his Ph.D. in 2002. In 2003, Dr. Hiller joined the RECARO Group: at first in the Holding as Head of Synergy Management, later he headed Customer Affairs at RECARO Aircraft Seating. In 2007, Dr. Hiller was promoted to the executive board as Chief Operations Officer. In April 2012, Dr. Mark Hiller was appointed as Chief Executive Officer of RECARO Aircraft Seating. In 2014, he also became shareholder of the company.

In 2004 he was honored with the Otto-Kienzle commemorative coin during the Hanover Colloquium "securing production sites through process chains" by Prof. Dr.-Ing., Dr.-Ing. e.h. mult. Hans Peter Wiendahl.

Joachim Ley was appointed Executive Vice President Supply Chain at RECARO Aircraft Seating, in 2012. In his current position, he is overseeing all supply chain related functions, in particular purchasing and operations across all RECARO Aircraft Seating sites.

He joined the Holding Company of the former Keiper Recaro Group in 2004 as Synergy Manager. Two years later, he took responsibility for Logistics at RECARO Aircraft Seating. From 2007 until March 2012, he served as Director Operations at the company's headquarters in Germany. During this time, he successfully completed a lean transformation in his area of responsibility.

Together with his team, he has hosted best practice events with more than 1000 visitors and held numerous speeches at lean congresses over the last years.

Joachim Ley holds a degree in mechanical engineering and business administration ("Wirtschaftsingenieurwesen") from the Technical University in Kaiserslautern, Germany, as well as the Simon Fraser University, Vancouver, B.C., Canada.

Vertical Integration: Titanium Products for the Aircraft Industry

Oleg Leder

Abstract This article concerns the importance of vertical integrated structures, employing the VSMPO-AVISMA Corporation as an example of a vertically integrated company that also forms part of a vertically integrated structure in the aircraft industry, to which it supplies semi-finished titanium products. In order to gain basic insight into the meaning of a vertically integrated company, value chain concept, which underpins the principles on which vertical integration is fundamentally based, must be appreciated. The value chain concept defines the links between the various parts of a manufacturer's operations in a systematic way, which enables it to identify how to improve overall efficiency, and hence to create value for the customer. If the value it creates is greater than the cost of conducting its activities, the business is profitable.

1 Introduction

Nine major technical and economic activities were identified as creating value, and these value chain activities comprise two categories, which are referred to as primary and secondary by Porter (2008), in Fig. 1.

Primary activities are directly concerned with the manufacturing process, from resourcing raw materials to the point when they are received by the end user, whereas secondary activities are those that support the primary activities to be accomplished. Each primary activity requires human resources, technological and purchase input, and the management, finance and legal infrastructure to support the whole chain. For that reason, there are interdependent links between the activities, and firms able to optimize the efficiency of the links generate a cost effective overall working practice. The continuing development of technology enables firms to gather information to determine the impact of activities on one another to find ways of reducing cost. The value chain model can also include links to external

O. Leder (✉)
VSMPO, Verkhnaya Salda, Russia
e-mail: oleder@vsmpo-avisma.ru

Fig. 1 The value chain (Porter 2008)

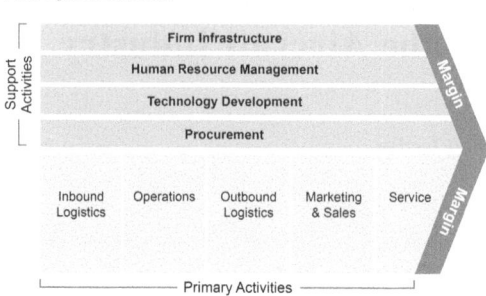

suppliers, which collaborate to reduce manufacturing costs, for instance, by supplying raw materials in a certain form, which reduces processing in the firm's value chain (Porter 2008). The longer value chain is referred to as the supply chain. Firms that create greatest value for clients have a substantial competitive advantage, so that strategies to continually improve the cost effectiveness of the supply chain represent a critical success factor for organizations. For that reason, the concept of a longer chain—the supply chain—formed by major organizations acquiring suppliers and integrating their activities into their own value chain, is called vertical integration. Although the value/supply chain is a simple notion, in reality it is frequently extremely complex. Cox (1999) compares it to a web with numerous inter-relationships between all the organizations that provide a highly diverse range of products and services to the main organization. Because there is substantial potential for inefficiency and poor cost control in a firm's supply chain, so that vertical integration of its suppliers enables the firm to have more control over costs and leveraging value (Harrigan 1984).

The rapid changes in technology and communication, which have occurred from the year 2000 onwards, drove substantial integration and unification in international relations along with economic, political and cultural globalization. The consequences for business were an ever increasing pace of change. There have been much fiercer competition between companies, a transition from the manufacturing to the knowledge age, requiring different skill sets and causing demand for high value added products (Kotter 2012; Bayliss et al. 2013). Vertical integration represents a major business strategy to reduce costs, to acquire new skills and to remain powerful and competitive in the market, often in a position resembling a monopoly; with consolidation as a major activity (TSBD 2002).

2 Theoretical Framework

The traditional economic definition of vertical integration, according to Buzzell (1983), it is a company in which two or more stages of production and/or distribution are conducted in-house, whereas they would usually be carried out in separate organizations. A whole industry can be a vertically integrated industry. Bonaccorsi and Giuri (2001) describe it as an industry in which a supplier can only sell to a downstream industry. The downstream industry cannot buy similar products from a competing sector. Hence a vertically integrated company within an industry is inferred as having an identical format, the manufacturing firm must purchase all parts and services from suppliers within its own company subsidiaries. It cannot purchase any product or services from suppliers outside the company. However, this is only one form of vertical integration, according to Harrigan (1984), a firm may not wish to conduct all operations itself and may be able to agree relationships with outside firms that minimize uncertainty, enable cost reduction without losing competitive intelligence. The strength of vertical integration is: recognizing which operations to perform in-house, how much ownership equity is involved in doing so, and when these factors should be modified to meet the challenges of new conditions.

Vertical integration occurs in two forms: (1) backward integration, when a company acquires suppliers that previously provided input to the business and (2) forward integration, when a company that had some part of dealing with its output are acquired (Johnson et al. 2013). Vertical integration can provide companies with a more efficient supply chain, leading to greater market power and minimizing the uncertainty represented by a firm not able to control its supplier's operations or the supplies to it (de Figueiredo and Silverman 2012).

2.1 Vertically Integrated Companies

A vertically integrated company is characterized by several features: it is a publicly traded company owned by its stockholders; it is an aggregate of companies that represent successive stages of one production cycle interconnected by means of their technological production relationship; it has access to diverse physical and human resources; its supporting product and service companies are enabling production development; its parent company has an effect on production operations and cash flow management (Kessler and Stern 1959; Johnson et al. 2013).

A vertically integrated company is a structure of integrated companies. It covers every stage of the production process, from sourcing the raw materials for production to the corporate entities converting them into components, from marketing to sales of the finished product. A vital component is innovation, it is an integral part of the entire operation. Vertically integrated companies potentially support a variety of economic purposes, for instance, cost reduction and access to economies of scale (Buzzell 1983). However, their common goal is to improve business activities, maximize profit, promoted by organizing and combining all production chain elements, and developing a common production and economic discipline for all production operations in the vertically integrated company (de Figueiredo and Silverman 2012; Porter 2008).

2.2 Advantages and Disadvantages of Vertically Integrated Structures

Establishing a vertically integrated company results in a new business environment, developing a system of inter-relations within the company, leading to a more effective general value system. As a consequence, motivation often increases, economic behavior and social responsibility of its participants grow, developing social awareness of participants and employees of a vertically integrated company. This is, however, dependent on the guidance and determination of previously autonom leaders within their area of operation (Sekkat 2006).

The positive and negative effects of establishing an integrated company, are summarized in Fig. 2.

The dependence on external companies to provide the company with products and services has a positive effect in terms of providing more certainty of the quality of operation and their timely performance. This dependence may also apply to raw materials, so that integrating such suppliers represents a fundamental need. Material shortages in companies go along with high fixed costs which are particularly damaging. Expensive facilities are not used to their full extend, so prices are raised and losses generated (Buzzell 1983). Improvement in production efficiency results from optimizing the production chain to reach production stability. In other words, the supply of finished products has to meet the demand (Hill and Jones 2009). The lack of dependence on external suppliers should also lead to a reduction in

Positive Effects	Negative Effects
Lower dependence on external supply	Lower levels of competition
Improvement in production efficiency and stability	Higher cost of management and control
Reduction in production and transaction costs	Lower flexibility
Increased production rate	Increased costs as outsourcing is no longer an option
Faster Return on Investment	Potential for corporate failure
Higher levels of innovation	Less focus on innovation

Fig. 2 Positive and negative effects of vertical integration

production costs, since the suppliers profit margins do not emerge (Harrigan 1984). This results in lower inventories, because the stock can be easily coordinated with production schedules, as suppliers are not involved (Buzzell 1983). Vertical integration can also diminish transaction costs that would have been incurred in activities, such as order processing, invoice payments and generating mandatory quality or legal documentation. The rate of production is also leveraged by means of internal process improvement, which is a continuous value-adding factor. The company can focus their investment decisions and gain Return on Investment (ROI) more quickly. A higher innovation degree is suggested as an advantage of vertical integration, because the business units participate in many production/distribution activities, and marketing as well as technical corporate functions are coordinated, rather than separated (Buzzell 1983; Albornoz et al. 2005).

The major disadvantages of vertical integration are: the competition level in the industry decreases, since competitors can no longer access components or services from the suppliers; a benefit for the acquirer but potentially a disadvantage for end users (Harrigan 1984). The firm's market power also creates high entry barriers for new firms (Buzzell 1983). Owing to the requirement for all business units in the vertically integrated structure to implement the centrally generated strategy, there is an increase in management cost and control; managers and control systems must be employed in each business unit (Harrigan 1984). There is often decreased flexibility, because the firm must use suppliers from its own organization, in other words the possibility of selecting suppliers does no longer exist. This may increase costs, as outsourcing may be much less expensive (Hill and Jones 2009), and may be less efficient in case the requisite skills are not present in-house. In addition, investment in a specific technology can be risky when the technology changes, especially in an environment of high technological change (Hill and Jones 2009; Buzzell 1983). If high investment is needed for vertical integration, profitability can be low unless costs are minimized. In some cases, the scale of operations may be too small for vertical integration to be efficient, usually it is most feasible for businesses with high market share (Buzzell 1983); this is also true if demand is unpredictable (Hill and Jones 2009). Excessive vertical integration can therefore be the cause of corporate failure (Stuckey and White 1993), for instance, owing to the focus on short term rewards, which restricts attention to future innovative directions (Buzzell 1983).

Vertical integration was considered a risky strategy by Stuckey and White (1993), who emphasized its complexity, the high investment requirement and the huge difficulty in reversing the process. These authors consider that many firms vertically integrate for the wrong reasons, and that there are four major reasons to do so:

1. The market is risky and unreliable
2. Companies at another vertical level have greater market power and are more profitable
3. Integration would increase market power
4. The market is immature

However, market risk is the most important issue, characterized by small numbers of buyers and sellers, and high asset specificity and intensity.

The quantitative study on vertically integrated companies conducted by Buzzell (1983) found that vertical integration is not profitable for all companies and that it is important

- to be cautious of high investment requirements,
- to identify and evaluate options for vertical integration, for instance, by agreeing on long term contracts that reduce transaction costs,
- either to integrate fully or in a minor way rather than incrementally,
- to implement only when there are large scale operations, and
- to ignore the potential for lower raw material costs unless the company has their own sources (Buzzell 1983).

2.2.1 Vertically Integrated Companies in the Aircraft Industry

The intensive development of aviation equipment, characterized by a new generation of commercial and transport aircraft each year, imposes new requirements for aircraft engines, operation and safety. Aircraft design is a complex engineering challenge, with an advanced tiered structure, diverse elements and large inter-communication networks; the airframe of a modern wide body aircraft consists of more than a million parts (Sadraey 2012). There are three major airworthiness standards in the industry: American (FAR), European (CS) and Russian (AR), although these are relatively similar, they retain some national features, and infer that each aircraft must be reliable and safe in operation. In the aircraft industry context, reliability is the capacity to maintain the required level of performance for a specific period of time, and flight safety to conduct flights without threat to human life and health (IAC 2016).

Aircraft manufacturers improve their competitive positions by becoming system integrators, and in recent decades, there has been significant consolidation owing to mergers and acquisition by the major players, for instance, Airbus reduced its supplier network from 200 when building the A380 to 90 for the A350 (OW 2015). The major vertically integrated aircraft construction companies support concurrently several aircraft producers.

3 Methodology

The design for this research combines explanatory and exploratory approaches since it seeks to identify links between cause and effect such as vertical integration and competitive advantage. It also embraces the human interpretation of events as perceived by individuals, which makes the outcomes uncertain owing to the different values and beliefs of humans that are based on their life experience, culture and social interactions (Saunders et al. 2012). For that reason, the research philosophy is pragmatist. Acceptable knowledge is recognized as arising from a variety of sources, multiple realities are the source of knowing. A pluralist approach to epistemology and ontology occurs respectively. The research strategy is archival, as all research evidence is sourced from records that have been prepared for some other purpose, but which are valuable since they provide a historical setting, reflecting change over a period of time, and focus on the everyday events of the company VSMPO-AVISMA based on research (Miles et al. 2013). The time horizon is a snapshot, since the research takes place over a short period of time and captures the situation at a specific time when the data is gathered.

The research methodology is qualitative, selected text is gathered and numerical data is limited to aspects of vertical integration, for instance, cost reduction and degree of consolidation in the sector. Secondary data is gathered from a variety of reliable sources, for instance, academic studies, company reports, trade magazines and quality newspapers as well as reliable websites, predominantly VSMPO-AVISMA. The sources of data are obtained using online databases, EBSCO and Google Scholar and Google search engine, by means of entering search words or phrases, for instance VSMPO, Airbus collaboration with VSMPO and vertical integration in the aircraft industry. The data gathered is read carefully to identify the studies that have relevance to answering the research problem. These studies are then re-examined for important words and phrases, which are coded by subjects. Once all the data is coded, the words and phrases with similar codes are combined and examined for themes, which can be interpreted by the researcher. These themes can then form a basis for analysis and discussion and finally compared with the concepts that comprised the literature review (Miles et al. 2013). Conclusions can be drawn by comparing study findings with existing theories on vertical integration and highlighting any new ideas, which were not previously reported in the literature.

The rigor of qualitative research relies on its replicability, transparency and credibility. In this research, all of the sources have been specified in the reference list, so that another researcher could repeat the study and gather similar findings. The transparency of the research, referred to as validity, is guaranteed by the degree of detail given in the findings and others are able to refer back to the original documents (Berg 2008).

In both selecting and documenting the findings, the researcher attempts to minimize own bias on the subject of vertical integration by using search words to find sources and attempting to interpret them as the original source intended (Berg 2008).

4 Findings and Analysis

This chapter covers the findings of the research on vertically integrated structures relating to a large vertically integrated company, and a supplier to the aircraft industry; *VSMPO AVISMA Corporation* (Fig. 3).

VSMPO-AVISMA (VA) Corporation is a vertically integrated company established in 1992, when Russia integrated into the global economy. Its Head Office is in Verkhnaya Salda, Russia and has subsidiaries in several European countries, USA and China. It produces high technology titanium products and its certified quality system was recognized as compliant with the world standards as of January 1, 2003 (VSMPO 2016a).

As a consequence of its vertically integrated structure, VA produces the required quality of titanium product from the ilmenite ore, which comprises mainly titanium and iron oxides. The processes include extracting the titanium by metallothermic reduction, separating the titanium sponge under vacuum, electrolysis and then forming, welding and machining the metal (VSMPO 2016b). It is the only vertically integrated company with the capacity for the whole titanium production cycle. It has leveraged its capacity to become a global player by its acquisitions, for instance, AVISMA in 1998, followed by GPK Titan in 2011, which has license for mining ore zirconium and titanium ores in the Tambov region in Russia (VSMPO 2011).

VA produces over 3000 titanium products, including large size forgings of disks and blades for aircraft engines, and complex shaped forgings for aviation landing gear, as well as structural forgings and other items. VA has gained certification by Airbus, Boeing, and Rolls Royce for its manufacture of sheets for the aircraft parts, which are produced by employing the superplastic forming technique. VA has longstanding partnerships with the world's leading aircraft and engine manufacturers, for instance, Airbus, Boeing, Embraer, Bombardier, Rolls Royce and Snecma. For many of these, VA is the major and/or strategic supplier (VA 2016c). VSMPO has a monopolistic position in the Russian titanium market. It is one of the world's largest titanium companies and deeply integrated into the global aerospace and engine building industries. VA's market share of the titanium products in the aerospace sector is approximately 30 %, and it has a 25 % share of the industrial market for titanium products. The Company participates in all major civil aircraft design projects launched by the leading aircraft and engine manufacturers, in Russia and globally. VA also produces large-size extruded parts from aluminum alloys and its aluminum division manufactures tubes, panels, sections and forgings for the aircraft and nuclear industries. The Corporation has a Research and Development Center with more than 500 professionals involved in scientific research. Its innovations and new technologies are based on a cooperation with Russian R&D establishments, as well as overseas partners (Fig. 4).

VA is an integral part of the vertical structure of large global aircraft companies, for example, the French company Figeac Aero. In 2014, they agreed on a long term partnership for machining titanium parts and manufacture subassemblies and assemblies for aerospace customers. As a consequence, a vertically integrated business structure developed. Jointly Figeac Aero and VA are able to ensure manufacturing process coordination and control throughout the entire production cycle, with the purpose of reducing landed costs for finished aircraft parts. VA's relationship with Airbus has developed since the late 1990s, when it signed first contracts with future key companies to form Airbus—Aerospatiale, DaimlerChrysler Aerospace and British Aerospace, followed by long term contracts in 2000s. This rapid-growing cooperation resulted in the most significant long-term-agreement (LTA), in 2009. It includes the supply of titanium to all Airbus divisions and sub-suppliers until 2020. The contract is representing more than 50 % of Airbus' titanium requirement.

The relationship between the two companies was described by DA (2011) as a strategic collaboration for manufacture and supply of value added products, not merely a joint development initiative, but a vertical integration project. The agreement is representing the strategic position of VA in the Airbus supply chain. The report also forecast that the collaboration would move from raw material purchase to rough and pre-machining of titanium parts by VA for Airbus, strengthening the degree of vertical integration. The CEO of VA stated that the partnership would assist in cost control and quality enhancement and noted Airbus' increasing dependence on VA for titanium. The Airbus Executive Vice President was also quoted as emphasizing the purpose of the partnership as a means of rationalizing and integrating its supply chain, to increase speed to market and to generate cost efficiencies (DA 2011).

In 2013, Airbus signed a Memorandum of Understanding with VA, which provides for a more extensive strategic partnership. It is described by Airbus (2013) as end-to-end collaboration in the areas of production, processing, and recycling of titanium products used in all Airbus aircraft programs, as well as the joint development of new titanium alloys for existing and future Airbus programs. On the same occasion Airbus and VSMPO-AVISMA signed a supplement to the 2009 contract, valid until 2020, for the supply of new forgings for the A320neo and A350-1000 aircraft.

The deepening integration of VA and Airbus is evident from another Memorandum of Understanding for the development of seamless titanium tubes plus related pre-material, which would be used in the hydraulic systems for all Airbus aircraft programs, and represented the strategy for product portfolio extension. The tubes were described as critical, high technology products for which few global manufacturers existed (VSMPO 2014). The relationship between the two companies was described as having significantly evolved over time to include the higher value products currently agreed on. It confirmed VA as Airbus' supplier of these vital raw materials (VSMPO 2014). As part of the long-term agreement with Airbus, the Corporation deals with more than 40 sub-contractors in the Group, the largest being PAG, Spirit Aerosystems, GKN, Aeroteam and Figeac Aero.

Fig. 3 VSMPO AVISMA plant

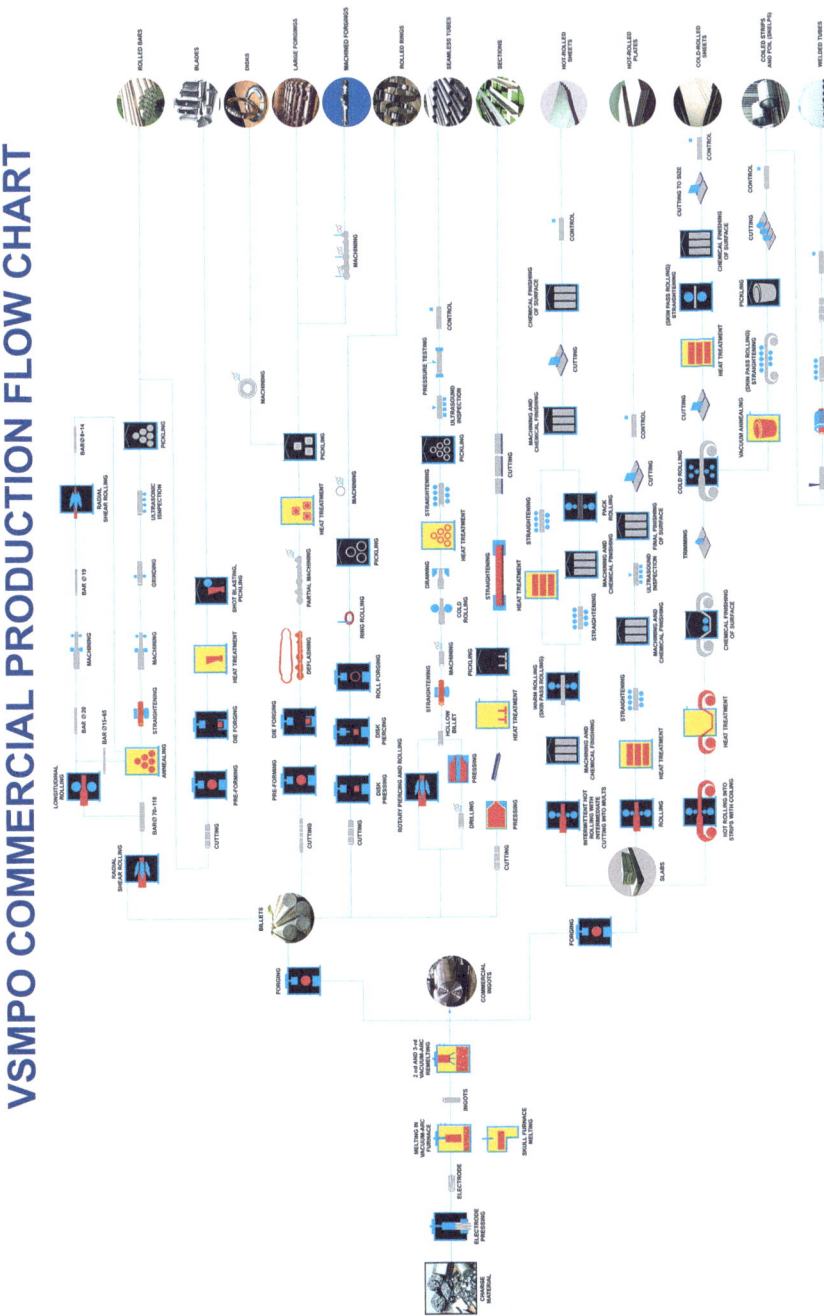

Fig. 4 VSMPO commercial production flow chart

5 Conclusion

This study has demonstrated that various forms of vertical integration can occur concurrently (Harrigan 1984). Although VA is a vertically integrated company, it is also part of a larger vertically integrated structure of large, more powerful companies within the aircraft industry (Stuckey and White 1993). This firm operates within the framework of industry vertical integration (Bonaccorsi and Giuri 2001), in which vertically integrated companies of various types exist, with some players such as Airbus decide on the amount of vertical integration they wish to participate in (Harrigan 1984). VA's vertical integration mode has been a backward integration in its acquisitions, but owing to the fact that its output are provided to companies such as Airbus, it could be considered that forward integration is also occurring (Johnson et al. 2013). Hence the complexity of the VA supply chain has significantly increased, resembling a growing web of inter-relationships as demonstrated further in these conclusions.

The acquisition of AVISMA, which produces the titanium sponge, along with emphasis on continuous improvement and R&D are the main drivers to transform VA from an ordinary raw material producer into a vertically integrated provider of titanium products. The crucial advantage of VA's vertical integration is: reduced dependency on external companies for various processes within its supply chain. One of the most important aspects is access to mineral ores containing titanium, by means of VA's acquisition of GPK Titan, in 2011. This reduced VA's exposure to fluctuating market prices (Buzzell 1983) and ensured raw material supply, so that it could operate a high fixed cost business at lower cost (Buzzell 1983). The position of VA was further strengthened and extended by its investment into capacity to undertake rough machining and by its partnership with Figeac Aero to supply finished aircraft parts. The supply chain of both companies intertwined and linked into the Airbus vertically integrated structure. Airbus is improving its competitive position by this introduction. Airbus has supported VA to improve its production efficiency, and to increase the rate of production of titanium parts for its needs, hence guaranteeing the supply of products it demands (Hill and Jones 2009). These activities demonstrate that the VA possesses are characteristically those of a vertically integrated structure. It is a major global company, and part of a tiered supply structure. It has access to human resources, technology and innovative development within the company, and by means of its contacts with external research and development facilities, aircraft manufacturers and other specialist companies (Kessler and Stern 1959).

As a vertically integrated company, VA has been able to leverage its competitive advantage by demonstrating the quality of its products and achieving international certification (IAC 2016) and aircraft industry approval by the largest manufacturers, providing the business with higher production stability and market power (Stuckey and White 1993). Airbus has used VA to ensure its supply of titanium raw material vital for its production, by agreeing long term contracts that provide cost and supply stability for the manufacturer (Hill and Jones 2009).

References

Airbus (2013) Confirms VSMPO's strategic position within Airbus supply chain. Airbus. Available via http://www.airbus.com/presscentre/pressreleases/press-release-detail/detail/airbus-and-vsmpo-avisma-to-reinforce-partnership/. Accessed 2 Jun 2016

Albornoz A, Milesi D, Yoguel G (2005) Knowledge circulation in vertically integrated production networks: cases of the Argentine automotive and iron and steel industries. Innovation 7 (2–3):200–221

Bayliss J, Smith S, Owens P (2013) The globalization of world politics: an introduction to international relations, 6th edn. Oxford University Press, Oxford

Berg BL (2008) Qualitative research methods for the social sciences, 7th edn. Pearson, Harlow

Bonaccorsi A, Giuri P (2001) The long-term evolution of vertically-related industries. Int J Ind Organ 19(7):1053–1083

Buzzell JD (1983) Is vertical integration profitable? Harv Bus Rev. Available via https://hbr.org/1983/01/is-vertical-integration-profitable. Accessed 4 Jun 2016

Cox A (1999) A research agenda for supply chain and business management thinking. Supply Chain Manag Int J 4(4):209–212

DA (2011) Airbus and VSMPO-AVISMA sign up to strategic collaboration. Defense Aerospace. Available via http://www.defense-aerospace.com/articles-view/release/3/128007/eurocopter-claims-russia-leadership,-delivers-ecureuils.html. Accessed 2 Jun 2016

de Figueiredo JM, Silverman BS (2012) Firm survival and industry evolution in vertically related populations. Manag Sci 58:1632–1650

Harrigan KR (1984) Formulating vertical integration strategies. Acad Manag Rev 9(4):638–652

Hill C, Jones G (2009) Strategic management theory: an integrated approach. South-Western Cenage Learning, Mason, OH

IAC (2016) Certification of aeronautical equipment and its manufacturers. Interstate Aviation Committee. Available via http://mak-iac.org/en/deyatelnost/. Accessed 4 Jun 2016

Johnson G, Whittington R, Scholes K, Agwin D, Regner P (2013) Exploring strategy text & cases, 10th edn. Pearson, Harlow

Kessler F, Stern RH (1959) Competition, contract and vertical integration. Yale Law J 69 (1):1–131

Kotter JP (2012) Leading change. Harvard Business School Press, Boston

Miles MB, Huberman MA, Saldana J (2013) Qualitative research methods, 3rd edn. Sage, Thousand Oaks, CA

Oliver Wyman (2015) Challenges for European aerospace suppliers. Oliver Wyman. Available via http://www.oliverwyman.com/content/dam/oliver-wyman/global/en/2015/mar/key-challenges-for-european-aerospace-suppliers.pdf. Accessed 23 Jun 2016

Porter ME (2008) On competition. Harvard Business School Press, Boston

Sadraey MH (2012) Aircraft design: a systems engineering approach. Wiley, Hoboken, NJ

Saunders M, Lewis P, Thornhill A (2012) Research methods for business students, 6th edn. Pearson, Harlow

Sekkat K (2006) Vertical relationships and the firm in the global economy. Edward Elgar Publishing, Cheltenham

Stuckey J, White D (1993) When and when not to vertically integrate. Sloan Manag Re 13, Spring 1993:71–83

TSBD (2002) The 5Cs: a framework for consolidation of fragmented industries. Tuck Business School, Dartmouth

VSMPO (2011) VSMPO-Avisma Corporation purchased Neftprominvest. VSMPO. Available via http://www.metalinfo.ru/en/news/50938k. Accessed 4 Jun 2016

VSMPO (2014) Airbus and VSMPO-AVISMA sign agreement on titanium seamless tubes. VSMPO. Available via http://www.vsmpo.ru/en/news/181/VSMPOAVISMA_rasshirit_nomenklaturu_titanovih_postavok_dlja_Airbus. Accessed 4 Jun 2016

VSMPO (2016a) In the world market. VSMPO. Available via http://www.vsmpo.ru/en/pages/V_mirovoj_rinok. Accessed 4 Jun 2016

VSMPO (2016b) Scheme of production. VSMPO. Available via http://www.vsmpo.ru/en/manufacture/Titan/sheme/Shema_proizvodstva. Accessed 4 Jun 2016

VSMPO (2016c) Quality and approvals. VSMPO. Available via http://www.vsmpo-tirus.com/quality-approvals. Accessed 4 Jun 2016

Oleg O. Leder was born in 1962 in Nizhniy Tagil. In 1985, he graduated from Ural Polytechnic Institute with a degree in Automated Information Processing and Management Systems.

In September 1994, he started his career at Verkhnyaya Salda Metallurgical Production Association (VSMPO) as Senior Electronic Engineer in Shop 10. In 2002, he was appointed to Marketing Director for Foreign and Domestic Market. Since 2009, he has held a position of Deputy Director General for Marketing and Sales at PSC VSMPO-AVISMA Corporation.

Part IV
Assembly and Integration

Quality Gates

Isabelle Sciannamea

Abstract Airbus wants to take measures to avoid any disruption causes in downstream aircraft assembly processes at its 11 production sites and four assembly lines. An increasing product design complexity, a global logistics network and aggressive ramp-up of new aircraft programs require an end-to-end strategy.

As part of a continuous effort to improve its quality processes and to mitigate disruption at production plants and final assembly lines, Airbus has decided to boost competitiveness by introducing end-to-end Quality Gates as a key element of the Quality Management System. This enables to ensure the maturity of deliverables all along Airbus' business processes, supporting industrial serial flow at all production and assembly steps.

From design to delivery: every time an internal or external supplier delivers a design, part, component or a complete aircraft, the customer will naturally check whether all specifications are met. If not, actions are taken by the supplier.

1 Introduction

In our private life, we all have high expectations as individual customers—we want to get what we ordered, what was confirmed and what we paid for. We believe that the need to apply the same high standards at work should be taken for granted.

If you're not a demanding customer in upstream assembly you might be creating a nightmare for the Final Assembly Lines (FALs) downstream. So we have to check that everything is as it should be at every stage. We have to strive for maturity levels that match general industrial norms, and not special Airbus ones.

The Quality Gate (QG) approach substantiates the decision to ship the deliverables from supplier to customer through a review of a set of measurable and initially agreed criteria. The Quality Gates contribute to secure and to improve Airbus

---

I. Sciannamea (✉)
Airbus, Blagnac Cedex, France
e-mail: isabelle.sciannamea@airbus.com

© Springer International Publishing AG 2017
K. Richter, J. Walther (eds.), *Supply Chain Integration Challenges in Commercial Aerospace*, DOI 10.1007/978-3-319-46155-7_13

operational processes through a regularly negotiated Customer Supplier Agreement.

It is a response to too much pending work prevailing all along the supply chain up to the point of use in the plants and FALs because of incomplete products from tier-X suppliers or internal production sites. This Quality Gate process enables visibility, a collaborative approach through agreement and support and has priority over cost reduction. It is a way to improve the on-time delivery, the quality and the maturity of the deliverables, and eradicates root causes for failed deliveries.

2 An End-to-End Scope of Application

The Quality Gate process is applied on the aerostructure, the cabin and on equipment critical work packages. Criticality can be driven by suppliers' non-performance, low process maturity and ramp up targets (new development, production rate increase, work package transfer, new supplier). Further stress comes from delivery process complexity or position on the aircraft assembly critical path, risk assessment and the impact of incorrect or incomplete deliverables. Experience from previous deliveries and lessons learned from other programs are essential.

The process applies to all Airbus civil programs, all types of deliverables (aircraft definition, set of drawings, set of specifications, a specific constituent or the complete aircraft), all design and assembly steps (from definition and design, supply, plants and FALs to industrial delivery across the whole Airbus serial industrial flow).

3 Definition, Intentions and Benefits

A Quality Gate is a set of measured criteria, initially agreed between supplier and customer providing the completion and maturity status of a deliverable and its quality level before delivery to the customer. All gates are actually part of an online flow chart that uses a red and green light system to make the interfaces between different process stages highly visible to management, especially to the programs (Fig. 1).

Based on the criteria measurements results shared during the so called Handover Meeting (HO) the customer is enabled to authorize or refuse the suppliers' delivery. When a red light appears, there is the need to start talking to the supplier, and record what is being discussed. The idea is to make the light sustainably green as fast as possible. Program management involvement can increase pressure on the supplier. It gives the supplier a chance to correct problems before the situation becomes out of hand.

Quality Gates

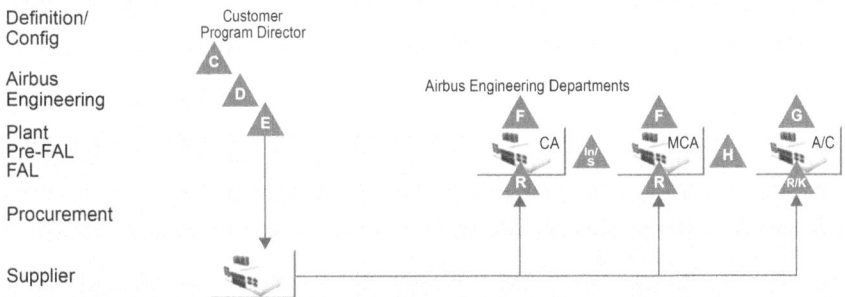

Fig. 1 Quality Gates all along the business process steps. Source: Airbus 2015

Many information can be derived from the process, among them: mandatory documentation, pending work, missing concessions, missing parts, status of completion, missing drawings, etc.

When a customer repeatedly experiences problems with the delivery quality, the supplier has to come up with sustainable solutions to get rid of the issue. This is the most important thing about Quality Gates. Once that customer-supplier discussion starts, there is a possibility of locating the root cause and solving the problem. Without this genuine, open dialogue nothing can change.

The Quality Gate process offers major advantages for management and for production workforce:

Efficiency and Performance
- Improve quality and maturity of the deliverables through clear customer/supplier relationship and transparency, Quality Gate results are disclosed to all relevant stakeholders and contributors.
- Smooth industrial flow along the supply chain and thus securing the aircraft program's ramp-up.
- Enable on-time and on-quality aircraft (AC) delivery to customers, reduce negative impacts on customers.
- Boost competitiveness, reduce costs thanks to a smooth industrial flow.
- Avoid snowball effects through solving issues as soon as they occur.

Standardization
- Strengthen the discipline by supporting shipment/delivery of deliverables at AIRBUS quality level through standardized methods and practices. Whatever the program, the phase or the location you are in, you run exactly the same process, with the same measures and reports.
- Harmonization at critical steps allows better understanding/visibility/transparency.
- Formalization of rules for consequential actions:
  Detailed recovery plans at Handover Meeting

Improvement loops for recurring issues, through the Practical Problem Solving Methodology (PPS)
- Handover meeting description to standardize delivery decisions.
- Harmonized measurement system with mandatory key criteria.

Mind-Set Changes
- Anticipation by providing information flows for early warnings.
- Pragmatic approach focusing on all the critical phases all along the production and assembly process.
- Strengthened supplier/customer relationship, clarification at all critical steps through involving key players all along the solution definition, formalized during the handover meeting and supported by a quality gate agreement.
- Adopting a challenging customer mind-set.
- Empower Airbus functions to accept or refuse (stop and fix) incomplete deliverables, employing a new mind-set of commitment to excellence and transparency.
- Encouraging continuous improvement, eradicate QG issues and meet a "tolerance zero" target.

The biggest benefit of all probably goes to the production workers in their contribution to final Airbus deliveries in the Final Assembly Lines (FALs). People do a fantastic job catching up with delivery targets but that can't remain the standard with rates going higher. Used in the right way, Quality Gates help to ensure that the work flow through the company to the FALs is smoother and that helps to make higher rates achievable.

3.1 Quality Gate Process Key Principles

Quality Gates (QG) are a set of criteria, initially agreed between supplier and customer, providing the completion status and maturity level of a deliverable before its shipment/delivery.

Handover meetings (HO) enable a customer to authorize or refuse the shipment/delivery of a deliverable. Suppliers present the completion status and maturity level of their deliverables to their customers. Based on this status, the customer authorizes or not (stop and fix) the shipment/delivery of the deliverable. This 'wake-up call' was used to create a shockwave upstream in case of an external supplier that repeatedly fails to deliver on maturity to one plant. On one shipset, Airbus refused to accept or pay for the sub-standard work. The problems were solved within 2 days. Since then, work that meets the specifications is delivered when expected. Pending work is accepted if it has no impact on the assembly process.

The QG criteria measurements evaluate the level of quality of the deliverable (complete and on-quality or incomplete). Criteria are defined in the agreement form with: name and description of the criteria, way of calculation/tool used to get the

Quality Gates

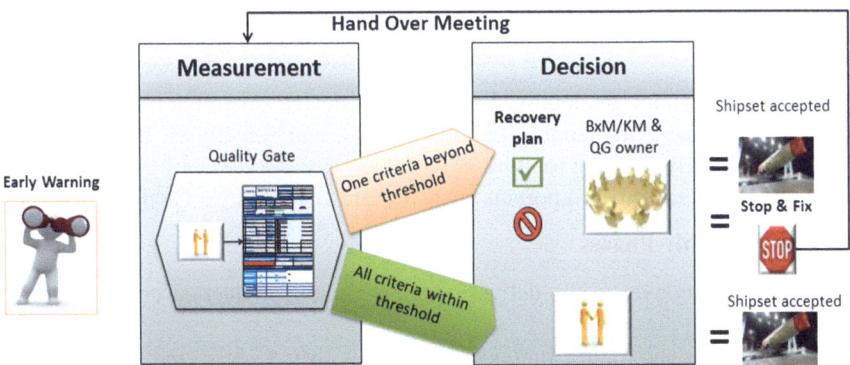

Fig. 2 The Quality Gate process, from measurement to decision. Source: Airbus 2015

data, threshold, criteria might require either numeric measurement or a statement (Yes/No).

The threshold indicates the tolerance of the criterion for the customer to accept or reject the delivery (such as maximum number of open non-conformities). The criteria thresholds reflect the capacity of the customer to manage an incomplete deliverable with the support of its supplier.

When the deliverable does not fulfil the expected criteria, the supplier informs his customer before the HO meeting (Early Warning) and provides a Recovery Action Plan until customer acceptance (Fig. 2).

Quality Gates ARE
- a transparent management decision tool
- a measure to support the shipment decision and lead to consequential management
- a check list with key criteria and associated agreed thresholds
- a completion state and the quality level of a deliverable before delivery
- a continuous improvement trigger

Quality Gates ARE NOT
- a contractual agreement on performance
- a (daily) pending work management tool
- a contract
- just paperwork
- substituting other directives or methods

3.2 A Quality Gate for Each Production Phase

Specific Quality Gates are deployed all along the design, sourcing, production and assembly processes, involving external suppliers, and sometimes, the suppliers' suppliers (see early warning toolkit).

Here you find the most important gates that are covered by a Quality Gate:

Aircraft Assembly Phases

L   Aircraft handover to the delivery centre, aircraft production completed and operationally tested
H   All sections and Main Component Assembly (MCA) delivered to FAL
O   Final Operational Test (FOT): aircraft completed as per definition dossier and system, tested on the ground. Aircraft ready for permit to fly issuance
S   Airbus inter-site deliveries for constituent assembly
IN   Internal delivery between plant and FAL, or between stations in FAL Sourcing Phases
R   External supplier deliveries to plants and FAL
K   External supplier deliveries of cabin equipment to FAL (customized equipment)

Aircraft Development or Modification Phases

CDCM   Preparation Cabin Definition Closure Meeting
C   Contractual Definition Freeze
D   Version Handover, transfer from Define to Design
EWO   Engineering Work Orders
E   Int Handover of all Interface points for Electrical System Installation (ESI) and Mechanical System Installation (MSI).
E   Customized Specifications Freeze
F   Section Installation Drawing Release
G   FAL Installation Drawing Release

3.3 Clear and Systematic Governance Principles

Two key actors are empowered by nomination and training to ensure effectiveness by process adherence: common priority setting and reactivity.

The Program Quality Manager ensures the consistency, sustainability and monitoring of the end-to-end (E2E) QG approach within a given program. He defines the Green Gate Plan according to program expectations with all key actors, ensures the realistic and ambitious targets and objectives settings for FAL Quality Gates and their consistent cascade. He supports the escalation and decision flow with the relevant stakeholders.

The Quality Gate owner (the customer), ensures the setup and effective running mode of implemented QG (agreed criteria and thresholds), runs the Quality Gate operationally, proposes stop and fix measures, ensures process adherence and drives green gate achievements (such as requests and follow-up Practical Problem Solving for red gates). He also ensures yearly agreement reviews, and decides on early warning tool application.

The escalation principles are important to mention. During the handover meeting, the supplier presents its recovery plan against a red gate. If the plan is not acceptable, in most cases because of critical impacts on the next gate, the Quality Gate owner informs its management who informs the program management, which can result in a stop and fix decision. In any case, when a gate is red, the next customer in the chain is immediately informed.

3.4 Key Behaviours in the Quality Gate Process

Needless to highlight that one major corner stone of the E2E Quality Gate process is the behaviour the process actors need to adopt. All actors play both a customer and a supplier role. Being a challenging customer is a key success factor to prevent a defect to be passed on to the next gate.

QG Customer
- I will take time to listen, check my understanding and act in a timely manner.
- I will ensure there is an owner for every action and a commitment to deliver.
- I will give constructive and transparent feedback.
- I will have the courage to say no, explain why and investigate alternatives.
- I will engage, challenge and support my supplier to solve problems, drive improvement.

QG Supplier
- I will ensure capabilities and means before delegating or making commitments.
- I will create transparency.
- I will ensure there is an owner for every action and a commitment to deliver.
- I will clarify and agree on customer expectations.
- I will engage, challenge and support my organization to solve QG problems/issues and drive improvement.

Programme QG Manager
- I will build strong teams across boundaries toward common objectives.
- I will engage, challenge and support my team to solve problems and drive improvement, encourage creativity and be open to new ideas.
- I will recognize contribution and celebrate success ensuring there is a balance with recognition, celebration and challenge.

- I will invest time with each team member to understand their needs and enable and support others to deliver and grow.
- I will have the courage to say no, explain why and investigate alternatives.

3.5 Quality Gate Implementation Steps

3.5.1 Define

The Quality Gate process set-up takes most of the time, and the added value is built through the definition of the Quality Gate criteria, thresholds and agreements. For external Quality Gates, this agreement supports the supplier to reach the contractual targets. One key success factor, for many topics of real business life, is to define a limited number of priorities, focusing on the critical deliveries. Thus, this first definition phase largely contributes to break the boundaries between functions, enabling a common comprehension of the business priorities (Fig. 3).

A Quality Gate agreement records the alignment between the customer and the supplier. It is a formal and binding document between the supplier and the purchaser, which defines the products subject to Quality Gate, the criteria and thresholds, the Way of Working (WoW) involving the timing for Quality Gate presentation and handover review organization as well as the communication

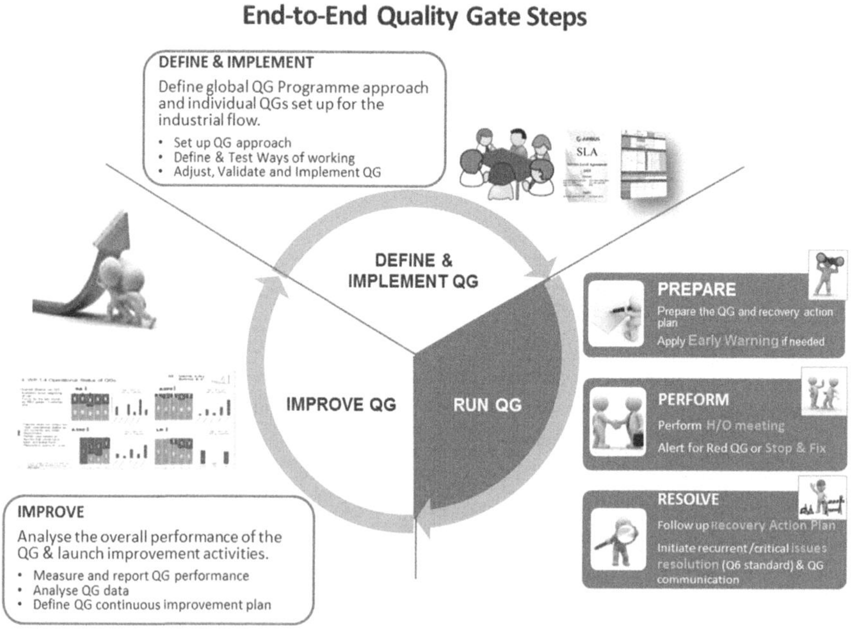

Fig. 3 Three major steps in Quality Gate implementation. Source: Airbus 2015

Quality Gates

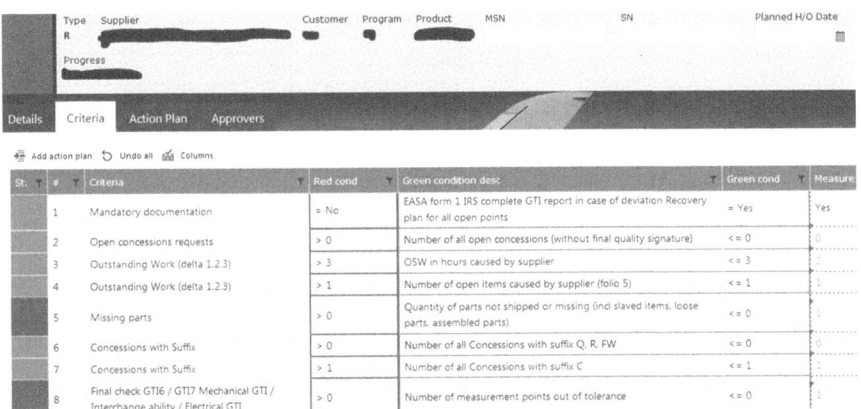

Fig. 4 Quality Gate assessment example. Source: Airbus 2015

matrix and the early warning features. Consequential management will monitor the completion of pending work in the Recovery Action Plan and associated actions to eradicate the problem (green/red consequences, communication and escalation). The Quality Gate Agreement should be a temporary measure with the objective of mitigating the risk of disruption to Airbus' plants and final assembly lines. In this phase, the Gates are defined with external suppliers, the contracts are established and signed.

Some criteria are systematically checked on Quality Gate milestones such as documentation, concession, pending work, missing parts. This can also be complemented by some other specific criteria such as geometrical interface acceptance, tests and inspection, non-conformities, etc. (Fig. 4).

3.5.2 Run

One key element of this phase is the HO meeting, which retro-plan versus delivery date depends on the program rate. The HO meeting, for example, is planned 2 days before the expected delivery date for A330 and A380, but half a day before for A320 SA (Fig. 5).

The supplier QG operation representative enters the QG data in the E2EQG IT tool with the measures for all the agreed criteria. He defines and fills in a Recovery Action Plan at least for one red criterion and/or a global Recovery Action Plan in case of several red criteria.

The QG supplier informs his customer on the clear reason of the red status and provides a targeted due date and an owner for each action (afterwards he communicates the real completion date).

In case of a red QG, the customer reviews the Recovery Action Plan. If the customer agrees with the Recovery Action Plan, it is "accepted" in the tool and the delivery is agreed. The status of the QG, however, remains red.

Fig. 5 Handover meeting agenda. Source: Airbus 2015

Agenda	
a. All Criteria Measured	☐
b. Quality Gate Status Reviewed *(criticality/impact of RED items...)*	☐
c. Recovery Action Plan set-up and reviewed	☐
d. Final Decision on shipment/delivery	☐
e. If Red QG, information flow [2]	☐
f. If **Stop&Fix** QG is proposed, escalation to appropriate level for final decision [2]	☐
g. If **Stop&Fix** QG is decided, information flow to appropriate level [2]	☐
h. Problem solving treatment launched for **Stop and Fix** *(PPS for internal airbus)*	☐

If the QG customer disagrees with the Recovery Action Plan, it is "refused" in the tool and the delivery is rejected. There is a stop and fix decision. The overall status of the QG remains red with a mention of "stop and fix".

In case of green QG, the shipment/delivery is automatically agreed by the QG customer. Nevertheless, the QG supplier can still offer a Recovery Action Plan if there are some open points.

Red QG status and/or stop and fix decisions require escalation. The QG results, the root cause of red issues, the Recovery Action Plan and the potential risk are immediately communicated during the HO meeting. This is a priority in case of a stop and fix shipment/delivery decision because the program management could supersede the decision. Escalation rules between the customer and the program are formalized in the standard agreement (Fig. 6).

Information/alert flow when a stop and fix decision is made with no impact on the next gate (Fig. 7)

One major objective in this phase is also to anticipate, as far upstream as possible, potential risks for a red QG in the delivery flow, to inform the customer of this situation and to mitigate those risks and be able to change the probability of a red QG to a green QG at the HO meeting and to define a reliable Recovery Action Plan as early as possible.

Because of the various types of QGs, and various criteria, different appropriate solutions are in place to enable alerts and to prevent red QGs (see Early Warning Chapter).

Detailed Recovery Action Plans at HOs and improvement loops for critical or recurrent issues are the key enablers for consequential management. Consequential management consists of monitoring the completion of pending work in the Recovery Action Plan and associated actions to eradicate the problem. If any risk is identified for incoming deliveries, a robust action plan (such as Practical Problem Solving) is implemented to eradicate/prevent the issue.

For Quality Gates with external suppliers, the Quality Gate owner is analyzing the Quality Gate results weekly, and if the trend is deteriorating, the escalation process starts, with a PDCA launched by the suppliers supply chain quality manager, at the end.

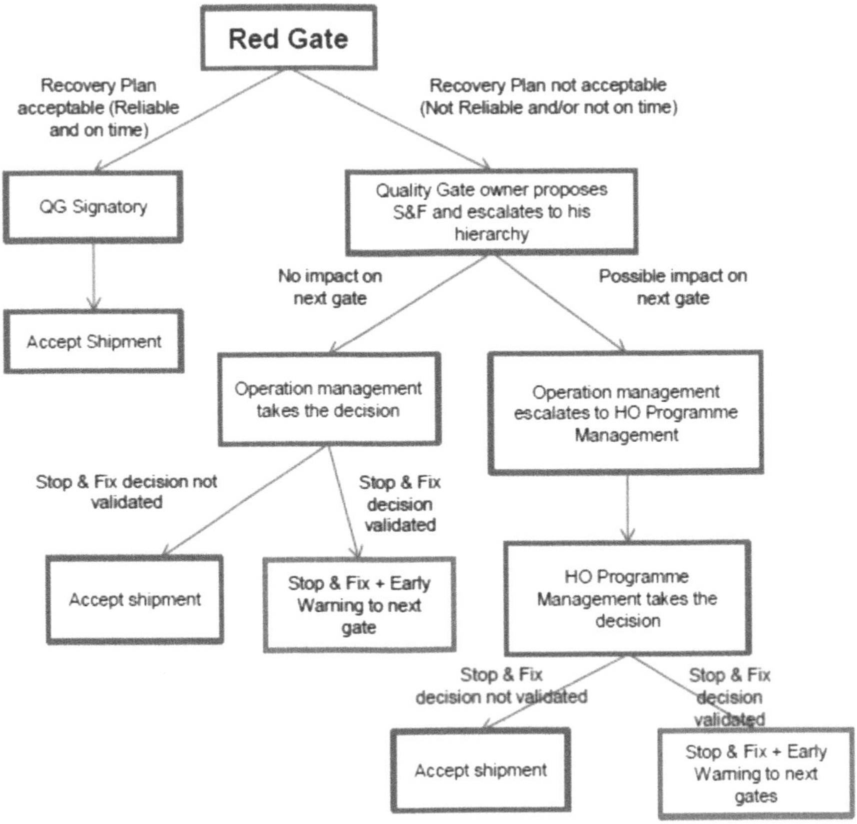

Quality Gates

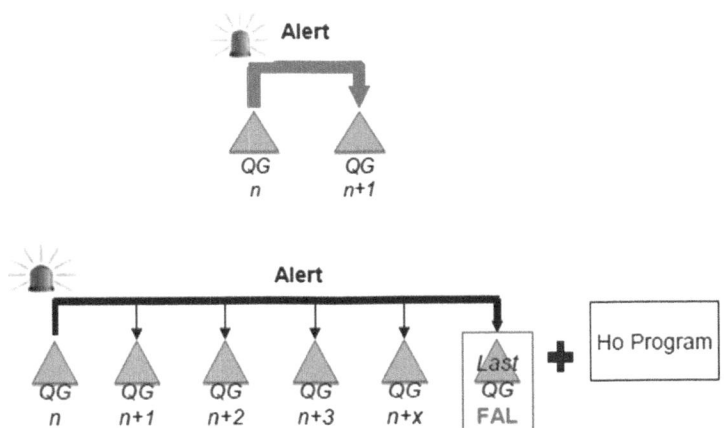

Fig. 6 Actions ensued from a red gate. Source: Airbus 2015

Fig. 7 Quality Gate escalation flow. Source: Airbus 2015

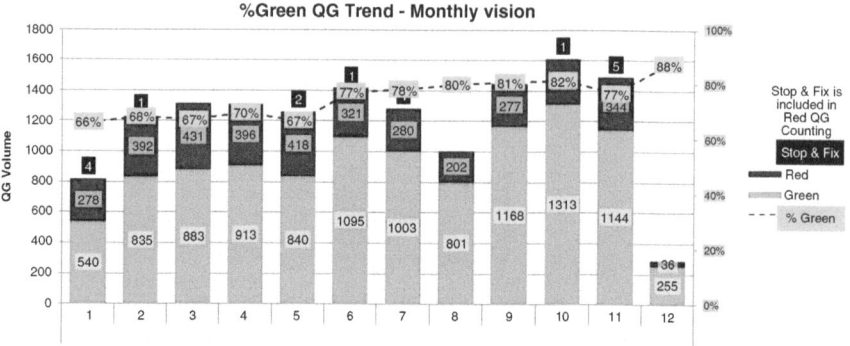

Fig. 8 Quality Gate performance chart. Source: Airbus 2015

3.5.3 Improve

This step drives the industrial improvement (Fig. 8).

Quality Gate Key Performance Indicators (KPIs) are taken into account of the Airbus top level Balanced Scorecard performance tool, controlling targets set by the program and cascading them backward all along the chain. These measures are used by the Program QG Manager and QG customer as the basis for Quality Gate results. The QG results are reviewed regularly during management meetings at FAL/Plant/Engineering/Procurement/Commodities level and also at program level. The main objective of these reviews is to sustain the QG methodology and continuously improve the QG performance.

4 Early Warning System

The objective here is to anticipate even more by being able to generate alerts in case of suspicion on a deliverable maturity, prior to the HO meeting. This prevents the potential of red QG risks in the delivery flow, enables to inform the customer of this situation and mitigates those risks as soon as possible. Changing and defining the probability of a red QG to a green QG at the HO meeting as early as possible requires a reliable recovery action plan.

Because of various QG types and various criteria, different appropriate solutions are aggregated in an Early Warning Tool Box which consists of six independent solutions. For each Quality Gate, one or several of these tools may be implemented, depending on the type of QG (internal or external), the nature of the supplier and the customer, and the type of program (high production rate or new development).

The customer decides to implement an Early Warning (EW) tool all along the end-to-end flow after discussing and agreeing this with its supplier, formalized in the Service Level Agreement (SLA) or the Quality Gate Agreement (QGA). The SLA is an internal Airbus contract, defining levels of service determining the what,

Table 1 Early warning system

	Delivery screening	– Analysis based on the next upcoming deliveries – Regular communication (weekly basis)	– Customer and supplier share and agree on the effects of specific and major risks – Customer can implement necessary actions (anticipation, priorities, etc.)
	Key criterion focus	– Tracking of one key criterion measured before the handover meeting – Timeframe allowing recovery actions on supplier side	– Communication by the supplier – Customer can decide about stop and fix or prepare recovery on-site
	Tier-2 Quality Gate	– Based on a Quality Gate set-up between external tier-2 and external tier-1 – Communication of tier-2 QG result done by the tier-1 to Airbus in any case	– Addresses the need to extend early warning to upstream
	E2E QGs map	– Share the Quality Gate status all along the value chain (previous QGs status allowing an anticipation upstream) – Generated out of the QG tool	– QG tool allows to analyze QGs status per MSN and per mandatory criteria – Available for all Airbus plants and FALs per MSN
	Immediate alert	– Based on the appearance of an unexpected issue, inform customer by using an immediate alert e-mail – Based on a standardized process	– Assign information to relevant person in charge (depending on who is affected) – Supplier provides root cause and impact study
	Pre-handover meeting	– Checking of all QG criteria (per MSN/deliverable) performed "x" days before the basic handover meeting – Check remaining activities to recover	– To keep green – Useful on low production rates or development programs

when and where a service is delivered. It is usually accompanied by the Key Performance Indicators (KPIs): (1) Handover governance, (2) Level of services expected from the customer and the supplier, early warning, (3) Level of performance of the criteria, (4) Way to monitor the performance criteria, (5) Improvement loops, (6) SLA and QG management and updates.

The major challenge is to choose the best tools that will help anticipate a red Quality Gate (Table 1):

5 Process Sustainability

The implementation of the E2E Quality Gate process was managed by the quality organization over a 2 years' time-frame. It entered service in June 2015. The major stakeholders were the programs, the procurement departments and the plants and FALs.

It was one of the seven key strategic initiatives followed at the Airbus Executive Committee level. The Quality Gate activities are now fully embedded in the different actors' jobs. A directive, a process and a method are referenced in the Airbus documentation and are available to all Airbus employees. An external version is available to external suppliers.

A specific tool has also been developed to support the E2E Quality Gate process. It is accessible by all actors, including external suppliers. Its user friendly interface and its simplicity were key factors for its adoption by all stakeholders. It is also a key asset to provide transparency.

A standard owner is nominated and responsible to maintain the process through the management of all documentation, manages IT tool updates, and associated training and awareness.

The quality team (Airbus Operating System) supports the effective deployment of Quality Gates and their sustainability, through standardization of ways of working across the company. The AOS team works with operational colleagues to develop effective routines that embed the Quality Gates approach into regular activities. They offer practical support on the ground to drive the new behaviours that can achieve the production targets. Both organizations constantly exchange feedback with the process owner to ensure that the system is tailored to meet the needs of different plants, suppliers and functions.

6 Success Stories

The wing team at Broughton plant embraced the idea of Quality Gates early and their customer in Bremen, head of wing equipping LR/A350 appreciates the E2E Quality Gate, too. Quality Gates make sure, there is no culture of blame or finger-pointing, just a strong customer/supplier relationship. Pending work and concessions are reduced and a powerful optimization boost can take place.

However, the system may face some limitations. It is indeed not so easy, to make stop and fix decisions when you have ten aircraft out of the production lines per week. Certain conditions need to be met such as having some buffer in the time-frame. In any case, the decision is always a trade-off with the rate, the ramp-up and the capacity of the organization.

Today, stop and fix decisions happen in only 2% of the cases but in all these cases, the impact has been huge. Decisions, however difficult they were, have turned out to be the right ones.

We can mention one Airbus supplier, for example, who was regularly delivering the work package to Airbus with missing documentation and in poor quality. This was leading to corrections and re-deliveries from the supplier, adding real burden to the St Nazaire plant, and having impacts on the FAL. Through the Quality Gate, Airbus decided not to continue to accept this poor delivery anymore, and the supplier was asked to stop the shipment, correct documentation issues at home, and deliver back to Airbus on-time and with the right level of quality. This, of course, led to financial implications, the supplier having now to pay for the disruption. It was a shock for the supplier but after 3 days review at their site, the next work package was delivered with the right level of quality. The Gate turned green, and it has stayed green ever since.

In some cases, when the risk to impact the Final Assembly Line is high, the escalation of the issue to the head of FAL is enough for the supplier to understand the criticality for Airbus and corrections are immediately implemented, so stop and fix is not used.

In terms of improvement areas, recurrence eradication is an element which requires constant effort and discipline, in particular for root-cause identification and resolution at external suppliers.

7 Conclusion

The E2E Quality Gate process in Airbus has implemented a real step change in the customer-supplier relationship all along the design, production, supply and assembly processes. Implementing the Quality Gate process on the A380 and A350 program was a real success. The transparency all along the steps is a clear asset to resolve problems as soon as they occur and prevent disruption upwards in the chain. It has now been deployed on all A/C Programmes. The process actually drives improvement. More and more green gates occur and suppliers understand the Airbus expectations better. And indeed, the Quality Gate process establishment was a key asset to reconciliate plant needs and procurement capabilities.

Isabelle Sciannamea, born 1969, started her career in Paris as Information System Engineer, working in real time development for Capgemini for 13 years. She moved to Toulouse in 1999, and became Senior Consultant in Capgemini, advising companies on information system alignment to their strategic objectives. In 2005, she joined Airbus. As cross functional Team Manager, she was working on aircraft equipment rejection causes, coordinating multi-functional teams. At the same time, she was a trainer at the Toulouse University of Social Science, providing courses on Information System design to Business Intelligence students in a Master course for 5 years. She was Head of the Strategy, Processes and Methods department in Airbus Supply Chain when she started this article. In May 2016, she was appointed Head of A330 Programme Procurement. Married, with two children, she is currently living in the South West of France, in Toulouse.

Lean Complexity Through Tailored Business Streams

Richard Hauser, Hans-Jörg Kutschera, and Benoit Romac

Abstract Aircraft manufacturers are always facing a trade-off between value of variety and cost of complexity. They must balance their customers' needs for aircrafts tailored to their precise needs, against the costs associated with the complexity of customization. Yet it is possible to gain both objectives.

Both complexity and costs can be reduced with a better understanding of customization. This does not mean eliminating individual features. Airlines always will need a high level of customization to remain competitive; product variations always will exist.

Instead, it means managing the choices of where and when to customize more effectively, and thus choosing which costs to regard as investments. The entire value chain—including suppliers, tier-x levels, and final assembly can be structured to allow for inherent variability. Parts, components, and process level can be segmented into three streams of activity: "basic and stable", "variable but predictable", and "random and emergent". These business streams can be tailored to the different needs of each customer.

This type of segmentation avoids the "one size fits all" approach that many aircraft manufacturers find themselves adapting to. After adopting that approach, one recent commercial aircraft client, for example, showed up to 30 % operating costs reduction along the supply chain. By dividing the value chain into three parts, the manufacturer ends up with a far more coherent whole system.

1 Value of Variety Versus Cost of Complexity

For an aircraft manufacturer, there is a trade-off between the **value that providing variety brings to customers** and the resultant **cost of complexity**. Airline customers, on one hand, want the cost benefits of scale and standardization. On the other hand, they value tailored products that help them to create a differentiated traveler experience and that optimize operations.

R. Hauser (✉) • H.-J. Kutschera • B. Romac
PwC Strategy&, Munich, Germany
e-mail: Richard.hauser@strategyand.de.pwc.com

Commercial aircrafts, both by definition and design, are highly complex products with deep and broad design content. Their bills of material often include more than 50,000 installation drawings and more than three million parts. While most elements of aircraft platforms (whether single aisle/short range or wide-body/long range) are highly similar −75 % of the aircraft does not vary from one airline to the next—the other 25 % varies a great deal. Customer requirements drive this variability, causing process issues across the full operational spectrum.

While trade-offs between standardization and tailored solutions are apparent, solutions are feasible. These solutions are only feasible, however, when it is apparent how the benefits of any particular customization will outweigh the cost. The cabin, for example, is a major element in defining an airline product. Airlines show individual preferences in the design of cabin amenities. They seek bespoke solutions for traveler-facing elements like seats, lavatories, IFE (in-flight entertainment) and even cabin lights. These are often considered branding elements for the airline. And they are comparatively inexpensive: while prices naturally play an important role in considerations of the cabin, it represents only 3–5 % of the total costs of an aircraft.

Thus, there is a great deal of leverage in investing in differentiation in cases like cabin, in which looking for optimal trade-offs between cost and individualization will not lead to an optimal solution. It is rather about optimizing operations along the entire supply chain as a way to minimize costs associated with variety along.

We call this strategic process "smart customization", or when it is combined with lean principles "lean complexity". As Fig. 1 shows conceptually, the profit impact of coping with complexity in a cost optimized way typically outweighs the impact of just finding the right trade-off.

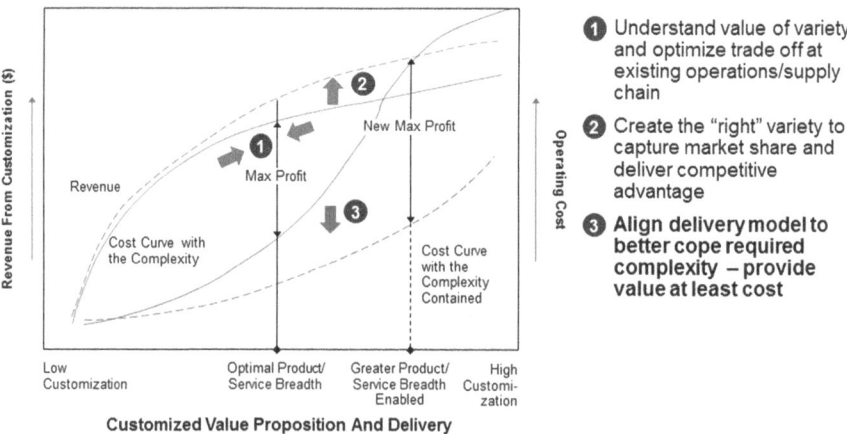

Fig. 1 Cost of complexity versus value of variety. Source: Strategy&

2 Understanding the Drivers of Complexity Costs

The best leverage for achieving lean complexity lies in product design. Modular product architecture—based on standardized platforms with interoperable components—allow for high variability on top of high levels of standardization. The automotive industry with its rigorous platform approach paved the way in this respect.

Modular architecture is a powerful support for customer backed variety. However, even modular architecture can create challenges along the supply chain, for example, at the points of maximum complexity in the final assembly lines.

In order to reduce costs without losing variability, a typical OEM view—a final assembly perspective—is not sufficient. Complexity hits different elements of the cost structure in different ways and in different places across the entire supply chain. Thus, an aggregated view across the total supply chain is required. As shown in Fig. 2, for example, material accounts for 85–90 % of the cost base in a typical assembly line. This number decreases to 10–15 % on an integrated view down to tier-2 level, and the cost base becomes most vulnerable to variations in product definitions.

It is important to understand what kind of variations drive complexity and which costs they incur. In one case of smart customization in this industry, analysis was conducted of requested variations along an aircraft supply chain from the OEM (final assembly line) down to tier-2 level. The research found that about 40 % of the aggregated cost structure was directly linked to observed variability, including modification and configuration changes, schedule and sequence changes, and operations procedures or vendor changes. This insight, even if it has a certain level of uncertainty due to missing cost details, helps illuminate the importance of operational approaches in minimizing the impact of variations (see Fig. 3).

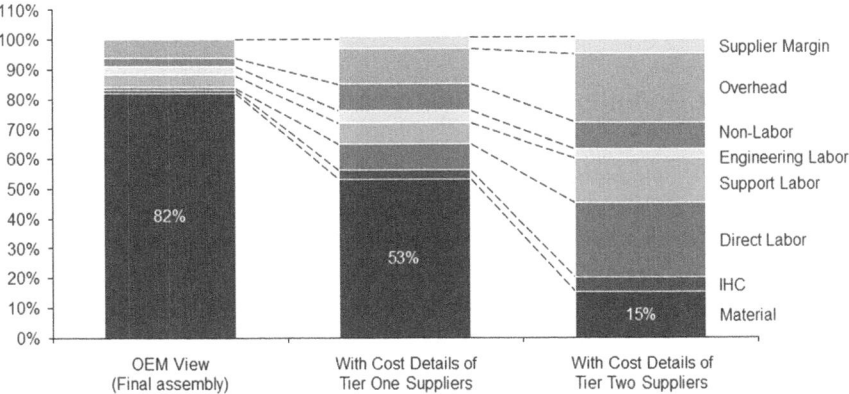

Fig. 2 Product cost structure across the supply chain. Source: Strategy&

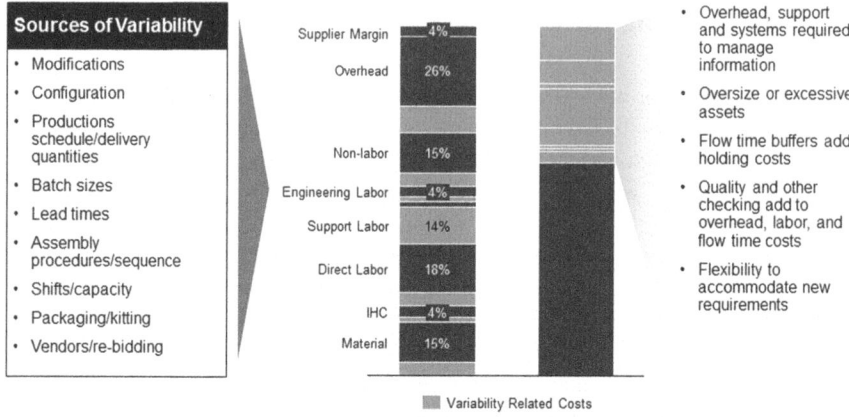

Fig. 3 Impact of variety on product costs. Source: Strategy&

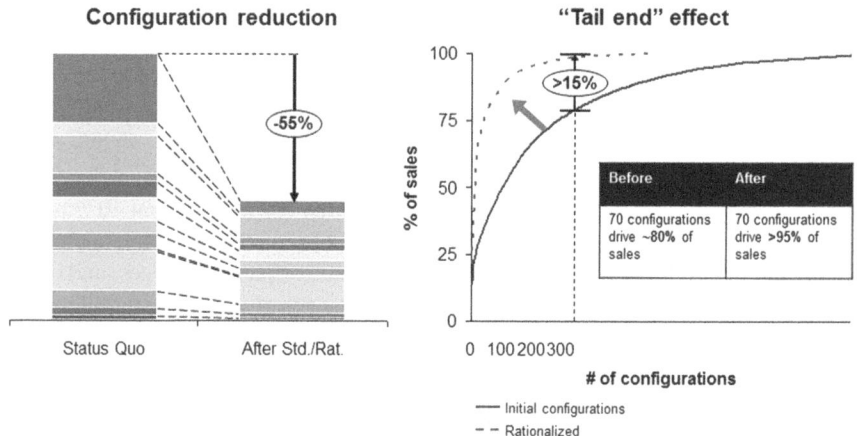

Fig. 4 Reducing complexity. Source: Strategy&

3 Limiting Complexity by Understanding Its Value

The first step in optimizing the costs of variety should be to avoid unnecessary variations and configuration complexity. Typically, the product development history includes considerations of a wide range of possible and actual configurations. Ad-hoc customer demands as well as incremental technology innovation and modifications often lead to incoherence: an array of design and configuration options, unrelated to each other.

But if you undertake a systematic review of the real value that these options create from a performance and application perspective, you can see how to create a substantial reduction at the tail end of the portfolio and increase the stability of the supply chain. In one case, analyzed in Fig. 4, this review led to a 55 % reduction in the complexity of aircraft interior component configurations.

4 Managing Complexity Through Smart Product Architecture

Having said that, the best leverage available for managing complexity is in the design of the product. When product design is maximized to balance variety and standardization, costs will be minimized.

For this, the architecture of the product is essential. A segmentation on component or parts level helps to differentiate between what is common and therefore stable against requirement changes (*"Runners"*), what shows a measurable albeit predictable variability (*"Repeaters"*) and what is really erratic and largely unpredictable (*"Strangers"*).

At an aircraft subassembly, we found that about 50 % of workload hours were dedicated to runners—components that crossed the line time and time again. Another 35 % were devoted to repeaters. Only about 15 % of deployed assembly hours were associated with strangers. But strangers had more of the costs associated with them than the other two groups. This finding supported the relevance of defining the degree of differentiation of that component to saving unnecessary costs.

The foundation of the core is by definition a stable runner. The "structurally different" parts are less stable. A modular concept can move a large portion of these parts to the *runners'* work stream, and an even larger portion to the repeaters, and thus cut costs.

This analysis, involving full differentiation of the components, should take place at the top level of the architecture. A product designed for managing complexity allows new parts and upgrades to be easily installed or applied in a "plug-in" mode. Depending on the full scope of required configuration flexibility, a fraction of these parts will unavoidably be strangers. However, if the proportion of strangers is limited, then the operational impact and generated complexity costs along the value chain, from manufacturing (batch sizes), to transport and stock management (buffering), and installation (missing parts, tool changes etc.) will be minimal (see Fig. 5).

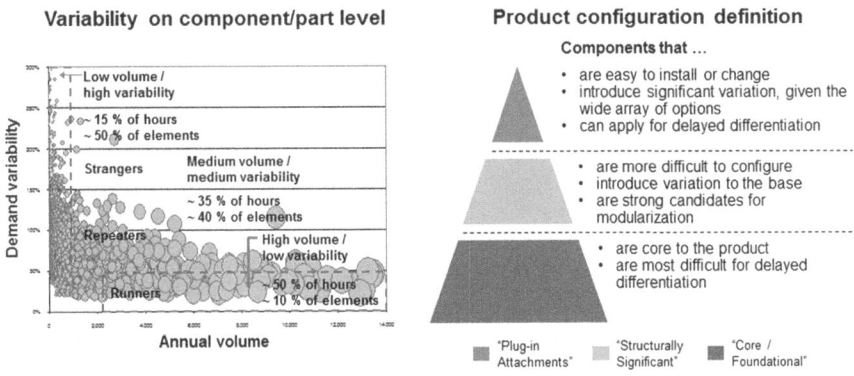

Fig. 5 Product variability and configuration. Source: Strategy &

5 Tailored Business Streams (TBS)

To be sure, a portion of variability cannot be eliminated through smart product architecture. That variability, however, can be isolated. You can segment the "runners" (those parts that are basic and stable) from the "repeaters" (that showed clear, predictable level of variations) from the "strangers" (those that are entirely random and unpredictable). You can then define the entire business process/stream along those three segments.

The initial driver for segmentation could be parts or components. In the previously cited case of smart customization, this analysis found that about 70 % of the parts, 87 % of the processes, and 94 % of the required resources (people, tools) were stable (see Fig. 6).

Before the re-structuring, the entire system was set up for the maximum level of complexity. Configuration and ordering systems were designed to manage variability, although 70 % of the bill of material did not need that functionality. Physical flows and assets were commingled, leading to worst case assumptions about back offs and buffers for the unknown. Compliance and tracking processes did not recognize distinctions. The weight of paper files exceeded the weight of the product as rolled out of the final assembly.

Then the operations were restructured along tailored business streams. The target was to define processes and policies tailored to specific characteristics and to define and apply differentiated targets.

(a) Basic and stable (Business Stream 1)
(b) Variable but predictable (Business Stream 2)
(c) Emergent and random (Business Stream 3)

For the basic and stable part (TBS1) the overall target was **efficiency and unit cost reduction**! The levers to achieve those targets included standard processes,

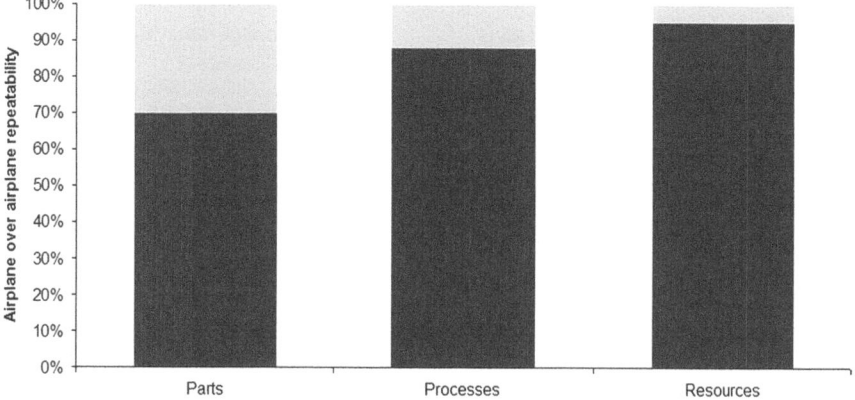

Fig. 6 Parts, process and resource stability. Source: Strategy&

automation, "first time right procedures", leverage of scale and low cost locations, aggregate forecasting, and capacity planning.

The overarching target for the variable but predictable part (TBS2) was **speed and reliability** in order to ensure high service levels. Supporting procedures included late stage customization in the assembly process, off-process set-ups, SMED (single minute exchange of die) technologies, pull based resource management, and worker flexibility.

The third business stream for unpredictable, random, emergent elements needed to be **flexible and cause as little disruption as possible**. For this, the process needed to be decoupled from the main stream and handled off-line as much as possible, ideally by cross functional teams.

The segmentation employed lean principles, with more focus and impact than a unified application across the entire scope of operations. Some of the analytical tools and methods used for each group are shown in Fig. 7.

Overall, the entire physical flow was designed in a more focused way, with dedicated flows, and balanced level load productions and resource consumption. Entire work stations were eliminated. Instead, parallel and easy control visual tools ensured smooth work flow.

A major impact of the TBS concept could be generated at the supply base. Before, suppliers had to cope with full variability, and at the same time, had limited or short-term information about changes. They had to invest heavily in flexibility though buffer stocks, excess capacity, red flag processes, and so on.

Now, the segmentation and decoupling of the demand along variability provided maximum stability and visibility to the suppliers. A more focused supply base emerged. Lean suppliers automated planning and scheduling interfaces to the OEM. Suppliers became flexible experts with tailored interfaces (see Fig. 8).

The implementation of the tailored business streams implied an evolution of the supply chain architecture to stabilize the demand throughout the chain. In the basic and stable business stream, suppliers were supplying the OEM in build-to-stock. That means they were not producing on order, but based on a long-term stable forecast. The stocks of components were held close to the final assembly line. In the variable but predictable business stream, suppliers were holding stocks of sub-components and producing components when receiving an order from the OEM. Finally, in the random and emergent stream, cross-functional teams were

TBS 1: Basic and stable	TBS 2: Variable but predictable	TBS 3: Emergent and random
• Synchronous delivery • Rate based planning • Self-inspection • Bottleneck flow reduction • Cellular manufacturing • Multi skilled teams	• Kanban /JIT • Pull based scheduling • Set-up reduction (SMED) • Last stage customization • Poke Yoke	• MRP/manual planning • Dedicated support teams • Off line operations • Flexible equipment

Fig. 7 Building blocks of tailored business streams. Source: Strategy&

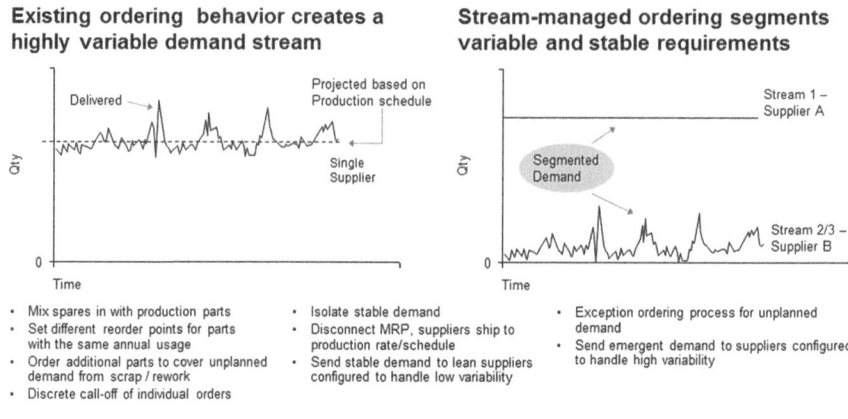

Fig. 8 Demand decoupling at supplier. Source: Strategy&

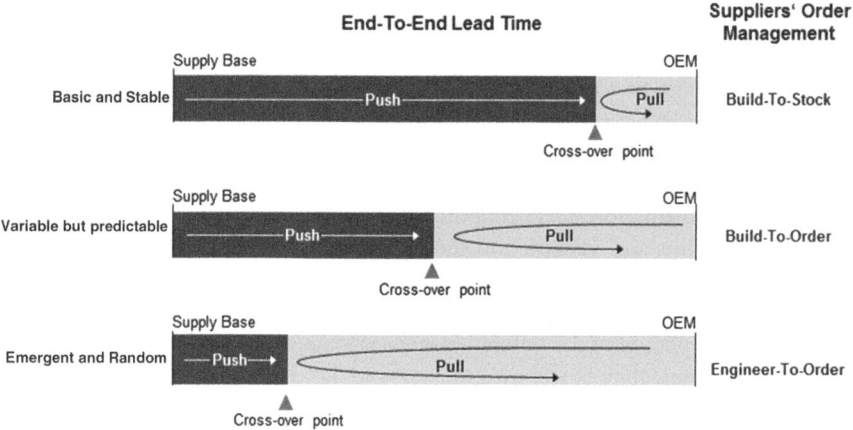

Fig. 9 Supply chain architecture. Source: Strategy&

involved to design the components with the objective of minimizing stock throughout the supply chain to avoid inventory carrying costs and obsolescence (Fig. 9).

Implementing the *Tailored Business Streams* concept required to align suppliers' lead times with differentiated lead times' targets. It helped stabilize the demand signal that was hitting the suppliers, while creating transparency across the end-to-end supply chain (Fig. 10).

These newly focused suppliers reported reducing the total cost base by 20 %, including saving up to 50 % of overhead (see Fig. 11 for one supplier's reported savings).

Lean Complexity Through Tailored Business Streams 217

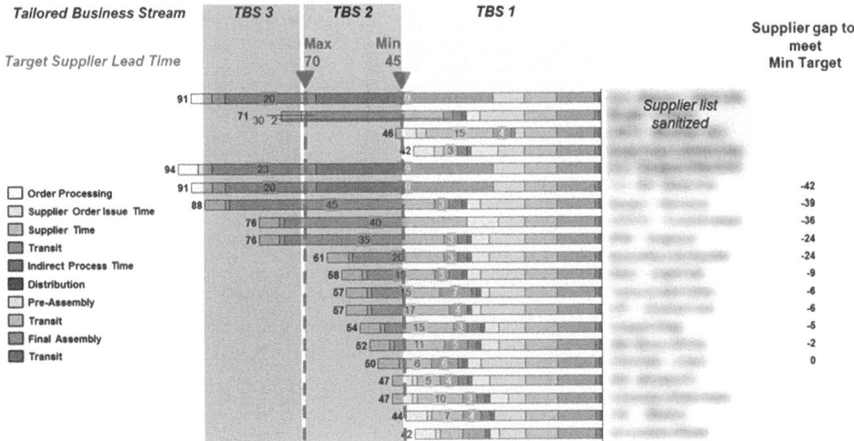

Fig. 10 Suppliers lead times. Source: Strategy&

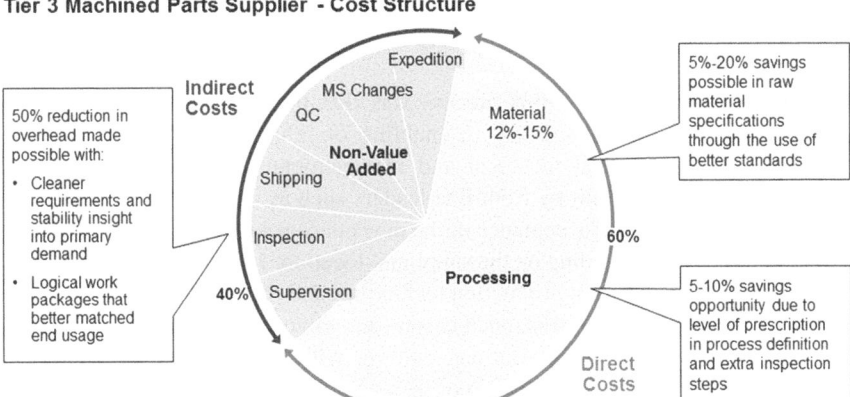

Fig. 11 Savings at suppliers. Source: Strategy&

Overall, significant improvements were achieved. In the aircraft assembly case, we managed up to a one third reduction of the total operating costs from OEM down to tier-3 level. Figure 12 shows the conceptual change. An initial "one size fits all" set-up, which by definition needed to cover that worst case, was transformed into a set-up of tailored streams with differentiated and tailored management approaches.

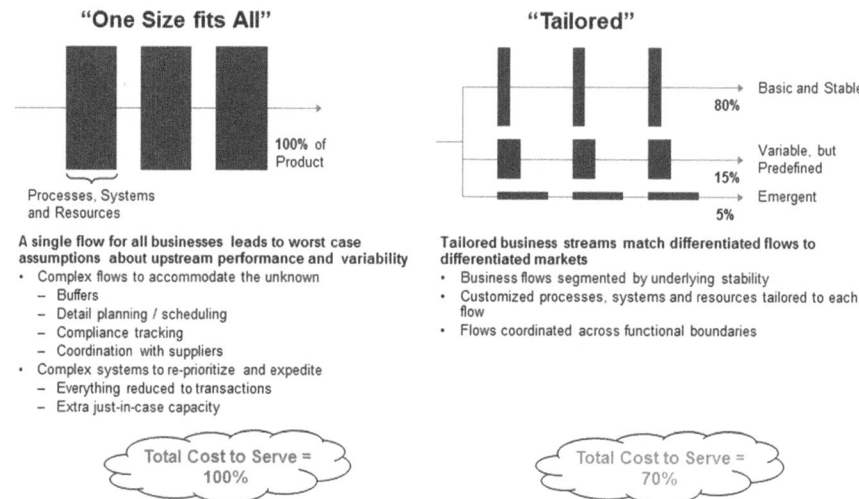

Fig. 12 Impact of tailored business streams. Source: Strategy&

6 Cultural Change

An important finding during this process was that the cultural background and attitude of the workforce was key to enabling or inhibiting change. Managing complexity with individual technical and management skills can be seen as a major personal achievement by front line leaders such as foremen.

We have found that the acceptance of the new concept varies across the different regions of the world depending on the suppliers' location. Japanese companies have long been working with lean production techniques and theories. Their approach is lean and standard by nature. It is much easier to involve workers in all ranks of the merits of the new concept. A German culture, with a high level of individual craftsmanship in its work ethic, may not be as amenable to change. Their culture is much more critical, and the critiques need to be considered and directly addressed. In the US, acceptance depends strongly on the level of qualification of the workers. Contrary to expectations, the more qualified and experienced workers are, they find it more difficult to accept the new concepts.

Nonetheless, any aircraft company can put practices in place like this. Your initial goal may be cost savings, but you may be surprised at the other benefits that occur: more satisfied customers, more of a reputation for customization, and an energized workforce.

Reference

Moeller L, Egol M, Martin K (2003) Smart customization: profitable growth through tailored business streams. Strategy + Business, Booz Allen & Hamilton, McLean, VA

Richard Hauser is a Vice President with Strategy& in the Munich office. He is an aerospace and aviation expert and has been a consultant to aerospace companies as well as for leading airlines around the world for many years. He is leading PwC's European Aerospace and Defense consulting practice.

Hans-Jörg Kutschera, Dr., is a Vice President with Strategy& in the Munich office. He is focusing on strategic and operational supply chain matters for Aerospace, Defense, and Security (ADS) as well as Industrial Manufacturing companies. His ADS clients are major airframe manufacturers and their supply base as well as defense equipment manufacturers. He is leading PwC's Industry 4.0/Industrial Internet of Things for ADS in Europe.

Benoit Romac is a Vice President with Strategy& in the Paris office. He is an expert in the aerospace and defense industry serving OEMs and suppliers. His focus is on operations and supply chain strategy. He is heading PwC's France Aerospace & Defense practice, as well as PwC's European cross-industry supply chain strategy practice.

Driving the Digital Enterprise in the Aerospace Industry

Helmuth Ludwig and Alastair Orchard

Abstract Digitalization is nothing less than a "question of survival" according to the German Chancellor Angela Merkel. For the aerospace industry, it is a huge opportunity, but it is also a vehicle for newer, more agile companies to leap ahead of existing market leaders, to realize their advantage and become Digital Enterprises today.

Technology is changing our lives. The way we travel, buy things, and interact with the world around us is different. Crowd sourced design, systems-driven product development, 3D printing, intelligent automation, advanced robotics and lifecycle analytics provide the technical foundation for a radical change to manufacturing industries. Some people even talk already about a fourth industrial revolution that will disrupt manufacturing as much as the digital camera did for 35 mm film.

Industries with the longest lead times and most complex products have the most to gain. History has shown that established market leaders are often the last to move and may be the most susceptible to agile competition—this is especially true in the digital world. A properly managed digital transformation should be at the top of the list for any manufacturer looking to be competitive in a highly sophisticated and demand driven world of 2020 and beyond (Fig. 1).

1 Introduction

Booking a holiday once involved a trip to the town center, flipping through printed brochures, a lot of sitting around while the travel agent typed cryptic codes into a proprietary terminal for a hefty commission. Today, we browse holidays online, use crowd sourced reviews to match a destination to our expectations and rely on an internet algorithm to choose the best deal. This transformation has radically

H. Ludwig • A. Orchard (✉)
Siemens, Munich, Germany
e-mail: helmuth.ludwig@siemens.com; alastair.orchard@siemens.com

Fig. 1 Internet technologies are increasing the pressure on manufacturers. Source: Siemens AG

changed our behaviors as consumers. It builds on pillars fully transferable to the industrial manufacturing world.

The first significant transformation driver is the digitalization of product and/or service companies. Digital products can be browsed in a global network. Connected customers consider transparency to correlate strongly with trustworthiness. A digital portfolio is also the pre-requisite for efficient customization, which is an example of newly emerging, often disruptive, digital business models that offer significant additional customer value. This is the second cornerstone of digitalization. Packaged holidays are out; personalized itineraries that cater for your specific needs are in. The third factor is the connected value chain. A Digital Enterprise can answer the question "How much would this customized product cost, and when can I have it?" more efficiently, and can subsequently manufacture it faster, at a lower cost and with higher quality than a traditionally organized company can.

Most commercial aerospace OEMs maintain a multi-year backlog, forcing airlines to make extremely long-term strategic plans that may prove risky as fuel prices fluctuate, global socioeconomic dynamics change and new legislation comes into effect. In the late twentieth century, with airline passenger traffic predicted to grow over 4 % annually during the next decade, airlines were eager to take delivery of newly designed, fuel efficient airplanes as quickly as possible. However, an airline ordering a Boeing 747-400 in 1990 could expect delivery of the airplane in 1997. Given this market environment, there were significant revenue opportunity benefits associated with shorter product flow time, and a clear competitive advantage to offer earlier product delivery (Chao and Graves 1997). However, backlogs continue to grow. Boeing booked gross orders for 1531 aircraft in 2013, delivering 648—less than 40 % to customers (Financial Times 2016). Airbus received 1036 orders in 2015 against an output of 635 jets (Gates 2016).

In common with any manufacturer looking for a competitive advantage in the digital age, aerospace OEMs need to address four fundamental challenges: reduce time to market, enhance flexibility, continue to exceed customer and aviation authority expectations in terms of product quality, increase efficiency and sustainability (Fig. 2).

Driving the Digital Enterprise in the Aerospace Industry

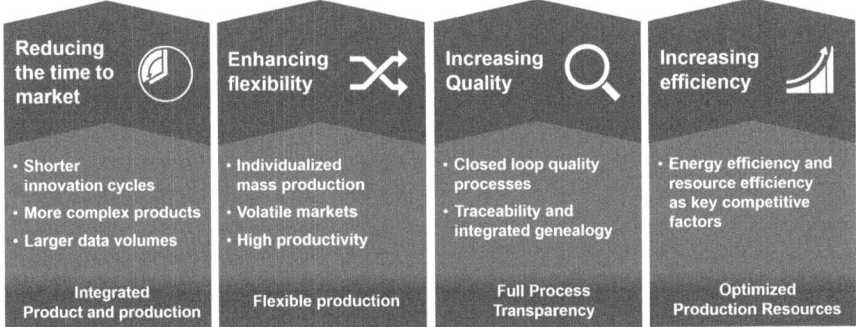

Fig. 2 The challenges manufacturers are facing. Source: Siemens AG

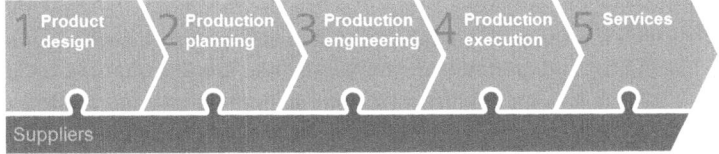

Fig. 3 Integrated product value chain. Source: Siemens AG

The enormous complexity of a commercial aircraft has led to pioneering use of digital technologies in structural design, guidance, navigation and control, instrumentation and in communication systems. Indeed, 3D CAD, which is core to all discrete manufacturing design today, was first presented as a vision in the book "Practical Analytic Geometry with Applications to Aircraft" (Liming 1944). However, the Digital Thread and the managed 'single source of truth' in one core "design and manufacturing model" which connects design, planning, engineering, production and service in industries such as automotive, is broken by the complexity of systems, processes, and cross company culture that make up most aerospace value chains.

Only an integrated approach guarantees long-term success. It should be the goal of all aircraft manufacturers to achieve a digitalized and integrated value chain—that manages all technical data in a seamless data model from design to service (Fig. 3).

This chapter looks more closely at the manufacturing related process steps. It will explain how a digital transformation can transform a manufacturer's internal value chain, to yield a sustainable competitive advantage.

2 Aerospace Production Processes

2.1 Parts Manufacturing

The materials used in aircraft production have dramatically changed since the wooden airframes of 100 years ago. In use today are steel, aluminum and other specialized materials such as lightweight metal alloys and composites. In both, commercial and military aircraft, aluminum, titanium and so-called super-alloys such as Inconel are widely utilized.

During the last decades, the focus was on optimizing subtractive procedures to achieve higher speed and quality. Advanced automated milling machines are able to efficiently recreate smooth part surfaces designed in CAD drawings to significantly reduce costs, compress the time from model to manufacture, and increase manufacturing flexibility. Traditional subtractive processes are now complimented and even replaced by adopting additive manufacturing—commonly known as 3D printing. The compressor inlet temperature sensor housing of GE's CFM Leap engine was the first 3D printed part certified by the F.A.A. to fly in a GE jet engine. Where 20 parts were once machined together, the entire housing is now a single piece that is five times stronger than the original GE design (Zaleski 2015). GE Aviation is now working to retrofit the new parts into more than 400 GE90-94B jet engines on Boeing 777 aircraft.

Together with the laying of composites that make up more than 40 % of modern aircraft, machining and printing are only possible through the combination of advanced CAD tools with sophisticated automation. Innovative motion control systems now allow the use of one type of common controller and one engineering tool for all such manufacturing operations, reducing the complexity of historically separated automation systems across the aerospace shop-floor, and contributing to the reduction and harmonization of cycle times across manufacturing facilities. Where part programs for machines were developed manually in the past, today the automation chain from Computer Aided Design to the part program is complete. The development of a part is fully automated, with a complete digital chain from part design to machining.

Utilizing these technologies, parts are programmed and simulated before they are transferred to the machine. This assures a highly productive machining strategy and up to 90 % increased availability of machining time during a changeover process (Fig. 4).

With the appropriate machines such as those from DMG Mori, manufacturers can combine Laser Deposition Welding (additive) and machining (subtractive) in order to produce previously unachievable design complexity with improved cooling features.

2.2 Structure Assembly

To deliver holistic improvements in speed, flexibility, quality and efficiency, part manufacturing automation has to be matched by similar advances on structure assembly; the sheer size of planes makes it a demanding task. If we go back to the origins of industrial aircraft manufacturing, then we picture "Rosie the Riveter" at work in factories during World War II with nothing more than a rivet gun (Wikipedia 2016).

Fig. 4 Virtual simulation of the motion control kinematics, virtual debugging of the motion control program and virtual start-up of the machine save time on the shop floor and improve manufacturing efficiency. Source: Premium Aerotec GmbH

Despite huge advances in the materials used, "Rosie's descendants" are still at work in aerospace today, aligning large stressed-skin elements, which distort under their own weight with tenth-of-a-millimeter accuracy.

Today, control systems that process Cartesian coordinates from a laser tracker support these tasks. This permits the precise positioning of machine axes and coordinated 3D path interpolation. Systems with over 90 axes have been implemented with negligible velocity losses using this technique, which was jointly developed between Airbus and Siemens.

It is clear that increased automation levels are required to eliminate the quality risks of highly manual processes. In industries such as automotive, where new car models are introduced each year, manufacturability considerations are already an integrated part of the more recent product design generations, allowing for efficient automation introduction that can be used and reused for millions of vehicles. In aerospace, completely new models are brought to market once in a generation, and costs must be absorbed by a relatively low output. Thus, products are still designed traditionally. Robots improve upon quality and efficiency of manual work whilst maintaining compatibility with existing assembly paradigms. They are the best way to achieve short-term gains.

2.3 Robotics

Drilling holes into components is the largest use of robotics in the aerospace industry. Precision requirements of this application makes robots the ideal means to quickly and consistently undertake this classically manual task.

While manual drilling may take up to four operations, robots drill the hole to its full diameter and depth, including the countersink, in a single pass. Robots now also paint, weld, seal, and inspect aircraft at a speed, and consistent accuracy level that humans cannot match. Critical to the productivity/cost equation, tooling costs are reduced when using robotics. Manual tasks require expensive tooling such as jigs and fixtures, whereas robots can operate effectively by relating their movements to a calibrated fixed point in space. This relaxes the constraints related to re-tooling costs, which combined with the increased speed and attractive learning curve, is a disruptive and cost-effective solution to increasing productivity. Using robotics, aircraft constructors are already claiming to achieve significant enhancements in monthly production. Lockheed Martin, for example, are moving from three F-35 aircraft produced per month to almost 12. In a market where deadlines are tight and backlogs are growing, robot integration is a must.

In addition to speed, the major advantage of robotics is quality, repeatability and consistency. Obtaining complete coverage on massive airframes, however, would require a dozen fixed robots. This would increase throughput, but they would not all run at full efficiency. Alternatives include fewer robots on servo tracks. They reduce manufacturing flexibility by inhibiting the reconfiguration of production lines. Mobile robots such as VALERI (Validation of Advanced, Collaborative Robotics for Industrial Applications) demonstrated successfully the use of collaborative, mobile robotics for applying sealant and conducting inspections in Airbus DS facilities in Sevilla in 2015, without the need for separating fences between the robot and human co-workers.

Taking mobile robotics a step further by combining highly mobile robotics with additive manufacturing is possible using groups of collaborating 'spiders' capable of 3D printing structures in situ. First prototypes (see Fig. 5) indicate enormous enhancements in speed and flexibility.

2.4 Simulation

Advances in the way robots are taught to perform tasks, is one of the reasons for the recent increase in their successful deployment. Once, highly specialized technical experts were required to program them, now robots can be trained just like humans. The graphical simulator environment calculates the most efficient movement for robots to complete tasks without impacting other equipment or endangering operators. In addition, robots can now adapt their motion and path based on the location and interaction with humans.

Instead of limiting the virtual world to products, 3D Manufacturing Engineering software now models robots, machines, logistics systems, and even people in a complete Digital Twin of the plant, the product and the processes. Rather than creating static pictures, the different elements can interact according to different scenarios until the machining and assembly operations are optimized (Fig. 6).

A significant benefit of a digitalized manufacturing environment should be, the ability to have "situational awareness" when considering changes. Since these Digital Twin models are economic to produce and are tied to the product design, tooling, robotics and plant, a thorough impact assessment of a considered change can be accurately made by virtual simulation. The simulations are used to verify

Driving the Digital Enterprise in the Aerospace Industry

Fig. 5 Fully autonomous additive manufacturing device from Siemens. Source: Siemens AG

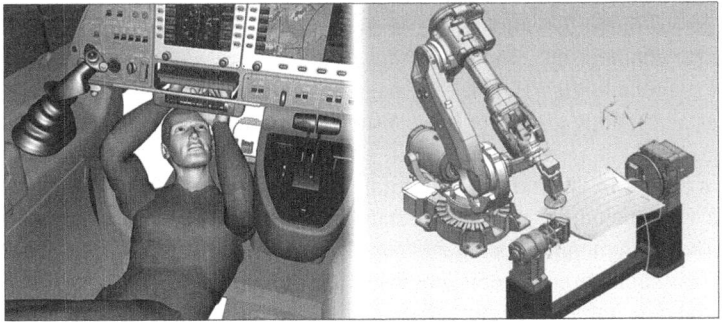

Fig. 6 Process simulation of manual assembly and robotic machining. Source: Siemens AG

feasibility, they can also generate important deliverables that extend the Digital Thread onto the manufacturing floor.

2.5 Work Instructions

Each manual simulation can provide a wealth of still and animated instructions to users, showing the safest and most efficient movements necessary to complete an operation.

The right operators get the Electronic Work Instructions (EWI) at the right time from the Manufacturing Execution System (see Sect. 2.8), removing any need to search for information and allowing staff to concentrate on productivity.

Being 'electronic' documents rather than paper-based means they can contain rich information and may be delivered through 'wearable' technologies such as Augmented Reality headsets—further enhancing the user experience. EWIs benefit from being generated on the fly, so they are always up-to-date and can reflect engineering or configuration changes that occur *after* the initial release to production—enhancing production flexibility and the agility of the company.

2.6 Automation Code Generation

Basic automation engineering is one of the points along the value chain where the Digital Thread is broken in most traditional enterprises, with engineers on the shop floor manually adjusting automation code in response to any design changes in the product. That certainly was the case in the past, but this paradigm is changing with the extension of the Digital Twin to cover automation. In this scenario, the code for the Programmable Logic Controllers (PLCs) can be automatically derived from machinery and robotics movements in the simulation.

Engineers can then test the automation in virtual PLCs against the simulated equipment in order to validate the accuracy and efficacy of their code, before pushing the changes out into the running factory.

Most manufacturers rely on external engineering companies to provide their machines and robotic assembly lines. Without the correct agreements in place, the construction of a complete Digital Twin is left to the OEM, and the additional effort is often a barrier to the company's digital transformation. Automotive OEMs have lead the way including automation standards, plug-and-play interfaces and full simulations in the tender documents for new lines. The suppliers are expected to deliver their simulations several months before ground is broken on the facility, in order to validate individual automation configurations using virtual commissioning, to check compatibility between production cells, and to ensure that logistics flow seamlessly through the entire plant. This approach is less common in aerospace, but it is an essential best practice to adopt.

2.7 Bill of Process Authoring

The Digital Twin is not only a teaching environment for robots and operators, it can also be used to examine and optimize every aspect of existing machining, assembly and logistics operations by applying Lean Manufacturing principles in the virtual world, before ever committing resources in the real world.

Lean manufacturing or lean production, often referred to as simply "lean", is a systematic method for the elimination of waste within a manufacturing system. As Womack, Jones and Roos explain in their classic 1991 book "The Machine That Changed the World", lean producers set their sights explicitly on perfection: continually declining costs, zero defects, zero inventories, and infinite product variety. The endless quest for perfection, on the part of lean producers, continues to generate surprising results. By combining continuous simulations to accurately model individual processes, and by discrete event simulations to capture the macro

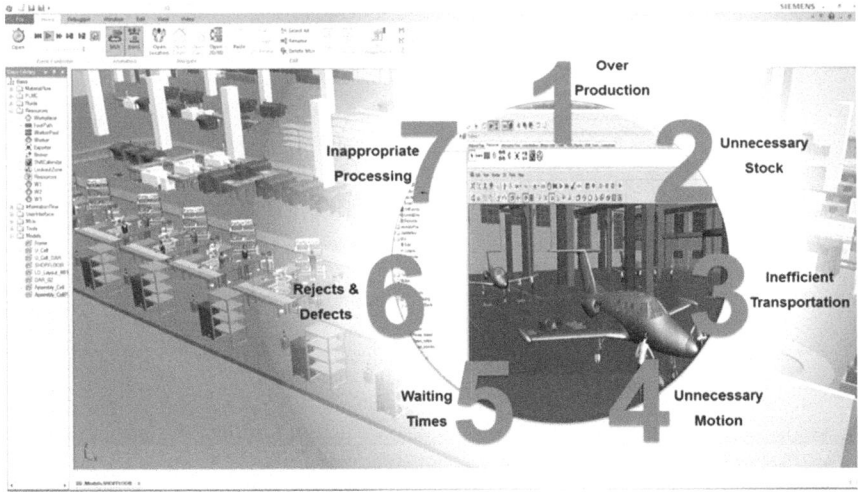

Fig. 7 Full plant simulation used to identify bottlenecks, balance lines and optimize workflows. Source: Siemens AG

behavior of the facility, manufacturing engineers can "look into the future" and proactively eliminate waste for any combination of planned activities (Fig. 7).

In order to reduce the effort of building a Digital Twin, drone-mounted laser scanners quickly capture a realistic representation of existing plant infrastructure. This is a tremendous advantage as many plants have existed for years or decades, and accurate 3D CAD models do not exist. The simulation, recognizes fixed elements such as walls, cables, and columns, and automatically incorporates them into the movements of the robots and operators.

70 % of manufacturing defects in aerospace are introduced at the design phase. Working in a virtual environment with the Digital Twin allows to "shift left" back along the value chain and resolve those problems before they impact manufacturing. The results of the simulated holistic production run is the Bill of Process (BOP) based on the engineering Bill of Materials (BOM) adding equipment constraints, part programs, tools, personnel and related work instructions, grouped per production step. The steps are chained together in a sequence, so that the simulation can be replicated in the real plant.

In "make to stock" industries, a single BOP may be associated to thousands of different orders that differ only by small variations such as color or packaging. These variations typically have a larger impact on logistics than engineering and are managed through the order handling process by the Enterprise Resource Planning system (ERP). The key relationships in those industries for technical data have historically been Product Lifecycle Management (PLM) <-> ERP and ERP <-> Manufacturing Execution System (MES) <-> Automation. For products that are engineered to order, each BOP is unique and can be considered the individual DNA of the complex configuration ordered by a specific customer. For this reason,

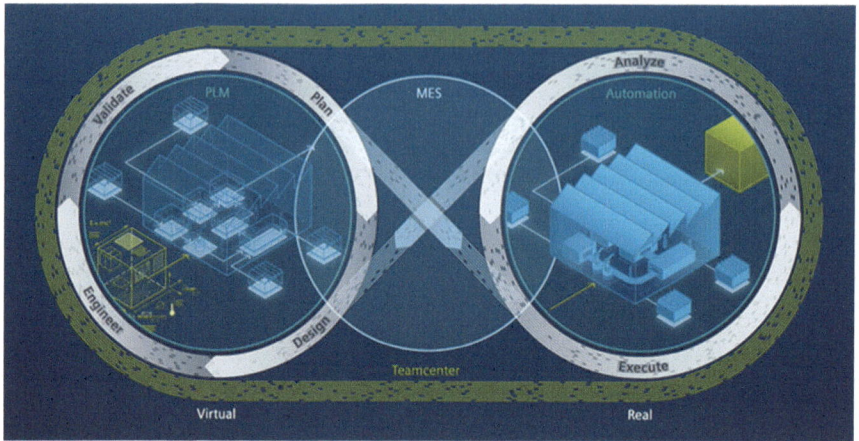

Fig. 8 Digital thread unifying the PLM, MES and automation domains in a seamless value chain. Source: Siemens AG

business processes cross continuously between the engineering and the shop-floor teams, and the key relationship for technical data needs to become PLM <-> MES <-> Automation with ERP focusing purely on logistics and transactional data (Fig. 8).

Such an agile connection between virtual and real world enables what is starting to become known as Closed Loop Manufacturing. A fully Digital Enterprise can respond quickly to customer requests by running an end-to-end simulation of the whole value chain—identifying technical feasibility of the request and defining which facility/resources to use for production. Order execution is more efficient because the whole process has already been optimized in the virtual world, and agility is improved to the point where partially assembled aircraft could be reallocated to higher-priority customers on the fly. This is a capability, some automotive manufacturers are already pioneering.

2.8 Bill of Process Execution

The picture of the factory-of-the-future is the one of smart products guiding themselves around a lights-out manufacturing facility, interacting with IOT-enabled devices, enacting the role defined for them during manufacturing planning. You can see the concept in action at Arena 2036 (**A**ctive **R**esearch **E**nvironment for the **N**ext Generation of **A**utomobiles)—a German project, where universities and companies collaborate to develop flexible factory concepts to build the car of the future.

This is how manufacturers expect to work 20 years from now, and governments around the world are working hand-in-hand with the private sector and universities to stimulate the technological, commercial and educational frameworks required to ensure their industries remain competitive. In Germany, there is Industry 4.0, in Italy the Fabrica Intelligente, then Smart Industry (Netherlands), Catapult (UK), l'Usine Nouvelle (France) and Made Different (Belgium).

The focus of each of these is on moving manufacturing towards technologies that are largely available today, rather than developing new ones. The huge investment, made in equipment and automation since the Third Industrial Revolution, is what potentially makes the next step a significant effort, requiring timelines measured in decades.

Roland Berger, the German Management Consultant, estimates that in order to assume a leading role in Industry 4.0, Europe will have to invest 90 billion euros a year over the next 15 years—a total of 1350 billion euros (Blanchet et al. 2014).

Just as the Automotive Industry has used cruise control, collision avoidance, lane-change warning, and other stepping-stone technologies to transition last-generation cars towards an autonomous future, so manufacturing requires a transition technology to begin unlocking the benefits of digitalization in the near term.

The ability of some existing MES platforms to coordinate and synchronize machines and operators actions combines perfectly with the Bill of Process to act as a proxy for the intelligent product, envisaged in forward-looking manufacturing paradigms. Key to such visionary concepts is that the PLM software, that authors the BOP, communicates on one common data model with the MES software that executes it, integrating both on a common collaboration platform that provides engineering and production uncompromised access to the Digital Twin.

Simulation is key to building manufacturability into product design; it is essential that manufacturers capitalize on this potential advantage by ensuring that the real production process faithfully follows the optimized simulations.

With the right MES and automation setup, the BOP can be used as a production blueprint to drive machining and assembly operations, and identify and resolve each deviation from plan. With these techniques, Siemens reduced lead times for the 1300-product portfolio in their own Amberg facility to just 24 h. Companies like Rolls Royce, IPT, and Maserati have seen similar increases in speed, agility and efficiency in their respective industries.

Aerospace assembly today is more similar to the Arena 2036 vision than it is to today's automotive manufacturing processes, and the introduction of BOP execution capabilities to the shop floor is already part of running projects such as Boeing's Fuselage Automated Upright Build (FAUB) project in Everett, USA (Automation World 2016).

A PLM-derived and MES-driven BOP can tie together islands of automation, coordinate intelligent robots with manual operations, call-out to automated logistics systems for just-in-time delivery of raw materials, check that every material and operation is performed as it was intended.

The Digital Thread ensures: requirements, established in the earliest phases of engineering, become features of the aircraft design; dimensions in the 3D models, simulated in the Digital Twin, published in the BOP, send to operators in EWIs and machines in part programs, are measured using sensors, and finally stored in the

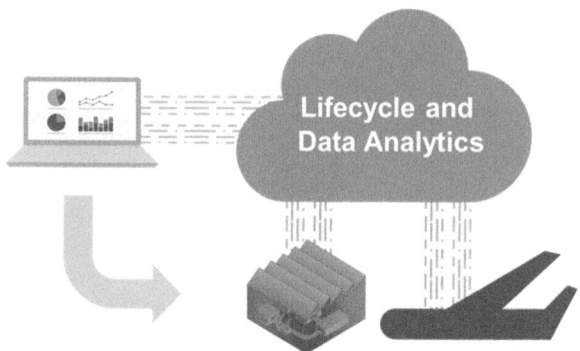

Fig. 9 Machine and product data analyzed in the cloud. Source: Siemens AG

as-built; the MES generates as the BOP is executed. The as-built record can be compared to the as-ordered configuration to highlight deviations on the shop floor versus customer expectations. Once published back to the Digital Twin it can be matched against the as-engineered definition as required by the FAA.

The as-built is also the starting point for the in-service part of the aircraft's lifecycle, and will remain linked to the Digital Thread and form the basis of the as-maintained record.

2.9 Data Analytics

The product, the process and the equipment in a Digital Enterprise all generate a constant data stream which can be captured, cleaned, contextualized and analyzed to derive real-time insights about a company's performance (Fig. 9).

The unexpected failure of mission-critical manufacturing equipment has such a negative impact on a company's ability to deliver product on time that most customers have opted to perform regular, preventative maintenance that lowers overall productivity by a real but predictable amount. The advent of cloud-based storage and computing has allowed sufficient processing power to be applied to large volumes of high-resolution machine data and recognize patterns which can predict failure in sufficient time to order and fit spare parts.

The result is what can be considered a "negative time to repair", and is valid to the reliability of products that are delivered to the end user as well as for machine uptime.

Analytics can also reveal patterns in underperformance throughout the value chain, driving continuous improvements in product design, process management, operator training, and customer engagement, and therefore represents a critical weapon in a manufacturer's armory—especially when combined with the foresight afforded by a continuously updated Digital Twin.

3 Conclusion

As demand for aircraft continues to grow, the aerospace industry has one of its largest opportunities to drive down existing backlogs, and increase competitiveness moving forward in integrating the manufacturing floor and Product Lifecycle Management into one common manufacturing model. As carriers look to benefit from a rapidly changing market, the aircraft suppliers who can combine safety and performance with the shortest time to market will have a significant edge over those who choose business as usual. To deliver this competitive advantage, companies need to integrate fully the Digital Twin and Digital Thread, providing them with the ability to lower production cost, improve first pass quality and increase production rates.

Other industries are already realizing the significant improvements in speed, flexibility and efficiency that digital manufacturing and automation can bring. Much of what has been subsequently built into the Digital Enterprise portfolios of some industrial software suppliers can be directly leveraged by aircraft manufacturers.

Digitalization is a huge opportunity for the aerospace industry, but it is also a vehicle for newer, more agile companies to leap ahead of existing market leaders. Previous developments have shown that once the tipping point for adoption of a disruptive technology is reached, change happens very quickly—and this is especially true in a digital world. Only 12 % of companies that were in the Fortune 500 in 1955 were still there after automation revolutionized manufacturing. By any measure, now is the time to start the journey to become a Digital Enterprise.

References

Automation World (2016) KUKA Systems to build robotic riveting systems for the Boeing 777 wide-body fuselage assembly. Available via http://www.automationworld.com/kuka-systems-build-robotic-riveting-systems-boeing-777-wide-body-fuselage-assembly. Accessed 22 Jun 2016

Blanchet M, Rinn T, Von Thaden G, De Thieulloy G (2014) Think act industry 4.0 the new industrial revolution how Europe will succeed. Roland Berger Strategy Consultants GmbH. Available via https://www.rolandberger.com/media/pdf/Roland_Berger_TAB_Industry_4_0_20140403.pdf. Accessed 22 Jun 2016

Chao JS, Graves SC (1997) Reducing flow time in aircraft manufacturing. Available via http://web.mit.edu/sgraves/www/papers/chaograves/chaograves.htm. Accessed 22 Jun 2016

Financial Times (2016) Airbus considers A320 ramp-up to match Boeing. Available via http://goo.gl/4NN4R6. Accessed 19 Jul 2016

Gates D (2016) Boeing trails in value of 2015 orders, but easily tops Airbus in delivery value. The Seattle Times. Published Jan 12, 2016. Available via http://www.seattletimes.com/business/boeing-aerospace/boeing-trails-in-value-of-2015-orders-but-in-deliveries-it-easily-outruns-airbus/. Accessed 22 Jun 2016

Liming RA (1944) Practical analytic geometry with applications to aircraft. Macmillan, New York

Wikipedia (2016) Rosie the riveter. Available via https://en.wikipedia.org/wiki/Rosie_the_Riveter. Accessed 22 Jun 2016
Womack JP, Jones DT, Roos D (1990) The machine that changed the world. Simon & Schuster, New York City
Zaleski A (2015) GE's bestselling jet engine makes 3-D printing a core component. Available via http://fortune.com/2015/03/05/ge-engine-3d-printing/. Accessed 22 Jun 2016

Dr. Helmuth Ludwig is Executive Vice President and Chief Digital Officer for Siemens PLM Software, a business unit of the Siemens Digital Factory Division.

Dr. Ludwig began his career at Siemens in 1990, working in Corporate Development to create regional strategies. He subsequently opened and built up the first Siemens organization in Kazakhstan, serving as General Manager until 1996, when he joined Siemens' Automation and Drives (A&D) group with responsibility for Process Automation Systems. From 1998 until 2001, he was Head of Siemens' Energy and Industry division in Buenos Aires, then from 2001 to 2007 served as Division President for Software and Systems House and A&D's Systems Engineering division. Dr. Ludwig served as President of Siemens PLM Software from 2007 until 2010, later taking on the role of CEO of the Siemens Industry Sector in Northamerica. In 2014, he returned to Siemens PLM Software.

Dr. Ludwig holds a Master of Science degree in Industrial Engineering from the University of Karlsruhe, a Master of Business Administration from the University of Chicago and a Ph.D. in Political Science from Christian-Albrechts-University in Kiel. As adjunct professor, he teaches for International Corporate Strategy at SMU's Cox School of Business in Dallas.

He is Chairman of the Board of the Commonwealth Center for Advanced Manufacturing and board member of the German American Chamber of Commerce—South. He serves as non-executive board member of Circor International, Burlington, MA.

Married with two children, Dr. Ludwig lives in the Dallas area.

Alastair Orchard is Vice President of the Digital Enterprise Project for Siemens PLM Software.

During his 20-year career in manufacturing, Alastair focused on the gains in operational efficiency delivered by Advanced Automation, Expert Systems, Manufacturing Operations Management, and Industry 4.0.

Alastair currently leads a global team running the Siemens Digital Enterprise Project, working with multinationals to become more flexible, cost effective, transparent, collaborative organizations, and to bring their excellent products to market faster than the competition.

Alastair holds a degree in Chemical Engineering from the Loughborough University of Technology in the UK and lives in Italy with his wife and four children.

Part V
Life Cycle Business Models and Aftermarket

The Aero-Engine Business Model: Rolls-Royce's Perspective

Peter Johnston

Abstract The aero-engine business has developed a business model quite distinct from that of the aircraft manufacturers. This chapter explains the fundamentals of this business, emphasizing how the differences in usage between aircraft structures and aero-engines lead to two quite distinct business models. The importance of lifetime in-service support to airlines is explained, as are the different service structures offered to operators. The innovation which enables Rolls-Royce to offer class-leading products and services is also described. It compares the aircraft and engine business models—and illustrates how the two models can operate together, using the newly developed A330neo aircraft and Trent 7000 engine as examples. Finally, we look into the future to further innovation in the business, highlighting technical developments which will reduce fuel burn and operating cost for airlines, and new service structures being developed to ensure engines can be supported in service through the different phases of their operating life.

1 Introduction

Rolls-Royce has been developing, producing and supporting gas-turbine engines for airlines since the late 1940s. During this time, the business model has matured to a point where engine services are as important to the business as original equipment sales.

The Rolls-Royce business model is to capture value from markets for high-performance power by developing advanced, integrated power and propulsion systems, providing long-term aftermarket support and delivering outstanding customer services.

The long-life products operate in challenging environments where they are expected to deliver sustained levels of differentiated performance. They deliver value to customers through outstanding power or other performance capabilities,

P. Johnston (✉)
Rolls-Royce, Derby, United Kingdom
e-mail: peter.johnston@rolls-royce.com

together with greater fuel efficiency and mission-critical reliability. This is often combined with a flexible service offering to best suit each customer's operating needs. This chapter will focus on the Trent engine family, which now includes seven engine types powering different models of wide-body airliner aircraft.

Significant investments in advanced technology and engineering programs are necessary to deliver market-leading products that recoup those investments. Many products have long production lives, while each individual engine may operate over several decades. Trent engines entered service in 1995, with the Trent 700 engine on the Airbus A330 aircraft. The latest engine to enter service is the Trent XWB-84, powering the Airbus A350, while the Trent 7000 is now in development and will power the Airbus A330neo from 2017.

Manufacturing becomes increasingly efficient as production levels rise, but high volume engine programs also secure strong aftermarket revenues. In certain markets customer relationships are strengthened through such long-term service agreements, aiming to deliver exceptional service standards and accepting risk transfer by committing to high levels of operational availability. This provides value to customers and in return long-term predictable revenues are achieved by the engine manufacturer.

By growing the installed base of engines over time, and consequently the aftermarket revenues they generate, the entire business grows, enhancing profit and cash flow. Cash flow is then re-invested to support future product development and technology programs to produce long-term competitiveness growth, with good shareholder returns.

Over time, aero-engines have become ever more efficient: they have enabled aircraft fuel burn improvements of over 70 % since the advent of jet airliners in the early 1950s (see Fig. 1). This has been achieved through unrelenting investment in technology (over $1 billion per year at present), and bringing technological innovations to the market such as the three shaft engine, the wide chord fan, and 'blisk' integrated compressor blades and discs. All these features are present on the latest engine, the Trent XWB, enabling it to become the world's most efficient jet engine.

Just as engine efficiency has been improved through innovation, so has reliability and durability, increasing the time each engine operates between major overhauls. Traditionally, suppliers would repair and service power plants whenever their condition required removal from the aircraft, billing operators for the servicing work performed at the overhaul. The resulting 'time and material' business model did not align the interests of engine manufacturers and airlines.

Rolls-Royce has deep, long-lasting relationships with airlines. When in the mid-1990s, a key customer revealed the difficulties that this traditional time and material business model caused, the company moved quickly to develop something new. The airline was suffering from unpredictable overhaul cost timing (it would have to pay for an engine overhaul whenever it occurred), and also unpredictable cost levels (it did not know how much each overhaul would cost). These two issues made airline financial planning difficult and risky—there had to be a better way.

In 1997, Rolls-Royce introduced TotalCare® with that customer—maintenance charged at a single price, simply calculated as a rate per engine flying hour.

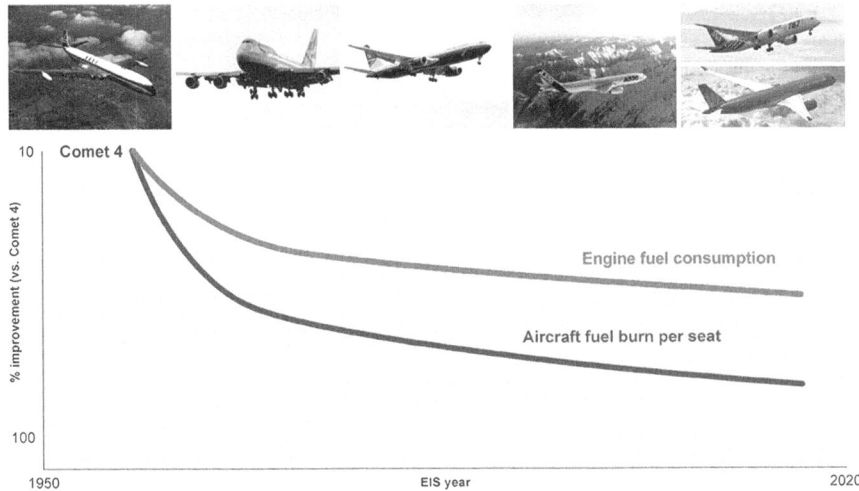

Fig. 1 Aircraft fuel burn improvement. Source: Rolls-Royce plc Marketing material, 2016

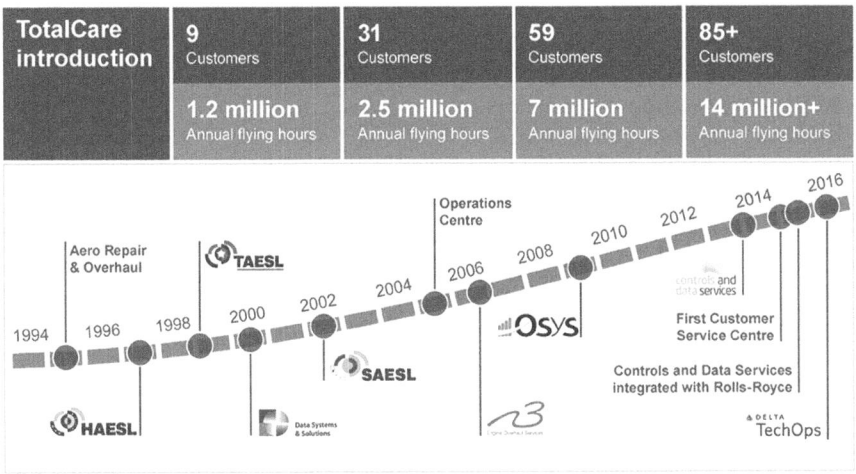

Fig. 2 Services—our journey. Source: Rolls-Royce plc, Services Marketing material, 2016

TotalCare solved both of the airline's issues, while also aligning the two companies' business models, and enabling all Rolls-Royce's knowledge and experience to be brought to bear in improving engine durability and reliability. It has become the foundation of the engine aftermarket and has been so popular that customers have chosen the service for over 90 % of Rolls-Royce Trent engines.

TotalCare services have since evolved (see Fig. 2) as airline operations have developed, incorporating new technology such as advanced Engine Health Monitoring (EHM), which enables Rolls-Royce experts to intervene to avert or minimize reliability issues. This shows the power of an aligned business model: under TotalCare, intervention to increase time between overhauls will improve financial returns for both the airline and the engine manufacturer.

2 The Business Model: Comparing Aircraft and Engines

Both civil airliners and their power plants operate under long-term, capital intensive business models, demanding the ultimate in technology. However, there are strong differentiators, because aircraft and engines operate in very different worlds.

2.1 Aircraft

The aircraft is a complex structure operating in a relatively benign environment. It will require maintenance to its subsystems and structure to ensure safety as flight loadings stress the structure and systems, using up their design lives—but this maintenance effort does not involve replacing major elements of the aircraft.

The business is characterized by high R&D investment over a 5–6 year period leading to certification and service entry. Production facilities will also require high investment over this period. Once in production, aircraft are priced to ensure a return on investment based on airline purchase alone, over an assumed production run. A successful aircraft will remain in production for 10–15 years before another major investment is required to replace or refresh the design. Each individual aircraft will remain in service for approximately 25 years, but over this period it contributes little extra to the aircraft manufacturer's revenue.

2.2 Engine

The engine operates with its key elements immersed in hot, corrosive gas streams, rotating at speeds sufficient to stress and wear individual components. Through its life, this demanding environment will require most high value parts in the engine to be replaced up to four times over. These parts are typically compressor and turbine blades, combustor parts, and life-limited discs and shafts.

The business model also requires high R&D investment over a similar period to the aircraft prior to its entry into service (known as the 'EIS' date). The investment, however, will be a fraction of that required on the aircraft. Due to intense levels of competition and airline buying power, engines are sometimes sold at levels approaching cost price, so new engine production alone does not generate a

Fig. 3 Cash life cycle of an engine program. Source: Rolls-Royce plc Services Marketing material, 2016

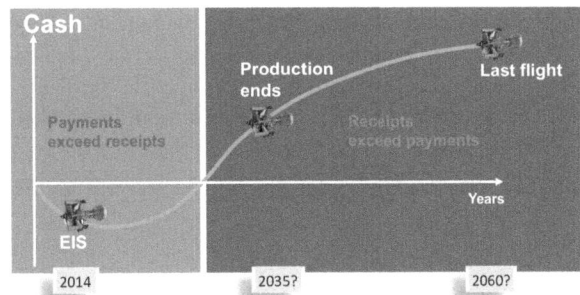

A long-term business: services revenue is the key to it

sufficient return on investment. Instead, the R&D investment together with profit required by shareholders is generated by the margins charged on the servicing each engine will require through its life. As airline customers are aware of aftermarket pricing at engine selection, pricing is constrained by competition, and engine and aftermarket form one linked financial system.

The cash life cycle is characterized by negative cash flow at the start of a program, and positive cash flow at the end (see Fig. 3) which can contribute to funding investment in the next generation of engines. Programs such as the RB211-535 (entering service in 1982) and the Trent 700 (EIS in 1995) are therefore important in funding the new generation Trent XWB power plant's development—and in turn, the Trent XWB will be expected to fund future generations of engine.

The Trent family has developed strongly since it entered service. The seven individual engine types within the family power wide-body aircraft were developed by both Airbus and Boeing, and while they share a basic architecture and design philosophy, new technology features have been incorporated as each new engine model has been introduced, in addition to growth in bypass ratio and increased core temperatures and pressures with each succeeding model. For example, the Trent 500 introduced three-dimensional aerodynamic design on key gas-path components, the Trent 900 introduced swept fan blades and contra-rotating core compressors, while the Trent XWB includes blisk (integrated blade and disc) compressor stages, modulated turbine cooling flows, and a new two stage intermediate pressure turbine architecture. As each engine has been developed, ever more sophisticated electronic control and health monitoring systems have been introduced.

2.3 Business Models Compared

Contrasting these two models, the airframe will demand higher up-front investment, which is recovered, together with profit, over a typically 15 year period. The engine will demand lower up-front investment and will receive lower revenues from sale of the engines themselves to airlines. However, returns will be spread over the complete production and service life: typically up to 40 years, made up of 15 years in production followed by 25 years operation for each engine produced. Adding the pre EIS R&D phase will therefore result in a 45 year long business cycle. This is a typical outcome—some engine program timespans will naturally be longer or shorter.

3 Business Model: The Outcomes

The natural outcome of the long-term aero-engine business is that manufacturers will seek to pull forward revenue wherever possible, as the 45 year business cycle results in a strong net present value effect on any early revenue.

The overall health of the business is fundamentally dependent upon service revenues, so engine manufacturers such as Rolls-Royce have invested heavily in innovative service concepts such as TotalCare, which create value for airlines while reducing the risk and volatility of revenues for the manufacturer. At Rolls-Royce, services now account for 53 % of revenue in the airline market—but a much higher proportion of profit. The aerospace market covers both commercial and military aircraft, but the commercial aerospace portion is growing faster than the defense sector—Civil Aerospace now accounts for 52 % of Rolls-Royce's total business. Therefore, overall services revenue on civil aircraft engines accounts for 27 % of Rolls-Royce's total revenue (all revenue values here are sourced from the 2015 Rolls-Royce plc annual report).

4 Services: Structuring the Aftermarket

Nearly 20 years ago, Rolls-Royce introduced the TotalCare service concept. This has been globally successful, with customers choosing this service structure for over 90 % of Trent engines. Under TotalCare, an airline pays a set fee for every hour flown by the engine. From this revenue, Rolls-Royce will pay for any service events (with some few exclusions for unusual occurrences). This structure has many advantages:

- The airline can budget for a predictable and consistent servicing expense, and is protected against the cost of unplanned events.
- Rolls-Royce is incentivised to work on keeping the engine flying and out of the overhaul shop, aligning its interest with that of the airline. Both entities will make more money when the engine is durable and reliable.

- Maintaining a large fleet of engines enables Rolls-Royce to invest in efficient, high technology, overhaul shops and health monitoring.
- The data collected from this large fleet of monitored engines (equivalent to receiving the content of Wikipedia every day) generates unrivalled knowledge and insights to make the engines more reliable and durable.
- Rolls-Royce receives a smoothed and predictable services cash flow, enabling planning and investment decisions to be taken rationally.
- The 'traditional' "time and material" model is still available for any customer unconvinced of the merits of TotalCare—but under this structure, incentives are misaligned: the engine manufacturer will make profit whenever engine maintenance is required, so arguably has less incentive to maximise reliability and durability.
- New service products such as TotalCare® Flex®, SelectCare™ and MRO Services are now being introduced to support airlines operating a variety of business models. These will be described in more detail below.

5 Optimizing the Business Model

The Rolls-Royce Trent family has now been in service for more than 20 years since the first engine was delivered, powering an Airbus A330, to Cathay Pacific Airlines. Initiatives to optimize the business model have not just focused on TotalCare, but also aimed at relationships with suppliers and with Airbus itself. Since the beginning of this relationship, Rolls-Royce Trent engines have been designed and developed to power the A340 (Trent 500), the A380 (Trent 900) and the A350 (Trent XWB). The culmination of this effort has been the agreement placing the latest Trent 7000 engine as the exclusive power plant for the A330neo aircraft. The market has responded, with the number of engines in service doubling over 20 years, and customers operating Trent engines all over the world—as shown in Fig. 4:

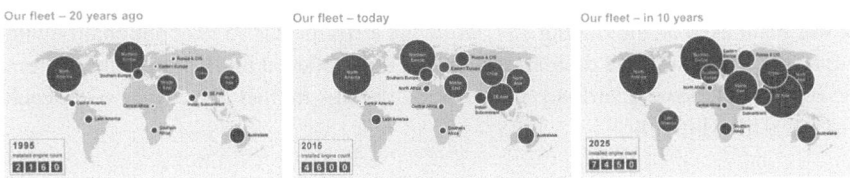

Fig. 4 Evolution of fleet sizes. Source: Rolls-Royce plc 2015 Annual Report, published April 2016

6 Suppliers: Risk and Revenue Sharing

A large proportion (over 70%) of the engine is produced by the external supply chain. The traditional business model here was for suppliers to simply build parts to a Rolls-Royce design, and be paid on delivery for them.

Through the evolution of the Trent engines, a better structure has been developed: Risk and Revenue Sharing Partners, or RRSPs. These are highly capable organizations, which have the resources and technology to design, develop and produce major modules or key components of the engine. Instead of being paid for producing parts, they provide a share of the overall investment required in the engine, and then receive a share of engine and aftermarket sales revenues. RRSP companies supplying major components for Trent engines include world-leaders such as KHI, MHI and GKN. Many RRSP companies supply several engines in the Trent family.

7 Working with Airbus: Innovation in Business Structure

The Airbus A330 has now been in service for over 20 years, and is the workhorse wide-body airliner for operators worldwide. It is available with engines produced by Rolls-Royce, General Electric or Pratt & Whitney—but the Rolls-Royce Trent 700 has been chosen for almost 60% of the A330s sold to date. The engine flies on more than 700 A330 aircraft, with over 70 operators across the world. The engine has completed over 38 million flying hours to date, generating detailed health monitoring information on each flight.

Having a choice of different types of engine for an aircraft type has often been popular with customers, as it gives them more negotiating power when dealing with engine manufacturers. It is, however, inefficient at an enterprise level: airlines are repaying three sets of engine R&D investment over the life of the program.

So, in 2014, when Airbus launched the A330neo program, it chose to power the aircraft exclusively with the Rolls-Royce Trent 7000 engine. This is an especially efficient project: the engine is a minimum change version of the Trent 1000 TEN power plant used on the Boeing 787, while the airframe has 95% commonality with the A330. The results are dramatic, with fuel burn reduced by 14% over the current A330 aircraft, saving airlines millions of dollars in fuel purchases over each aircraft's life (Fig. 5).

Fig. 5 Trent 7000 and A330neo. Source: Rolls-Royce plc (Trent 7000), Airbus SAS (A330neo)

8 Secure the Future

Airline expectations of the aero-engine industry continuously increase: in order to fly ever more passengers around the world, more efficiently, more reliably and with less environmental impact, these customers expect each new Trent engine design will not just provide ever-improved technology within the engine, but will also provide improved service products to support it in operation.

Analysts predict a continuing strong civil aviation market, with passenger traffic growing at an average of 4.6 % over the next 20 years. Aero-engine manufacturers are competing hard to share the market which this growth provides, with new products and services continually in development.

Rolls-Royce is therefore investing strongly in new products, with the Trent XWB-97 for the A350-1000 joining the Trent 7000 in service from 2017. Beyond these developments, the next engine generations are already being designed: the 'Advance' and 'UltraFan™' concepts (shown in Fig. 6) introduce new core engine technology, Carbon-Titanium (Cti) fan systems, and in the case of the UltraFan, geared fan drive, to the wide-body market. Engines based on these concepts will be available during the next decade, providing fuel efficiency up to 25 % better than the Trent 700 powering today's A330.

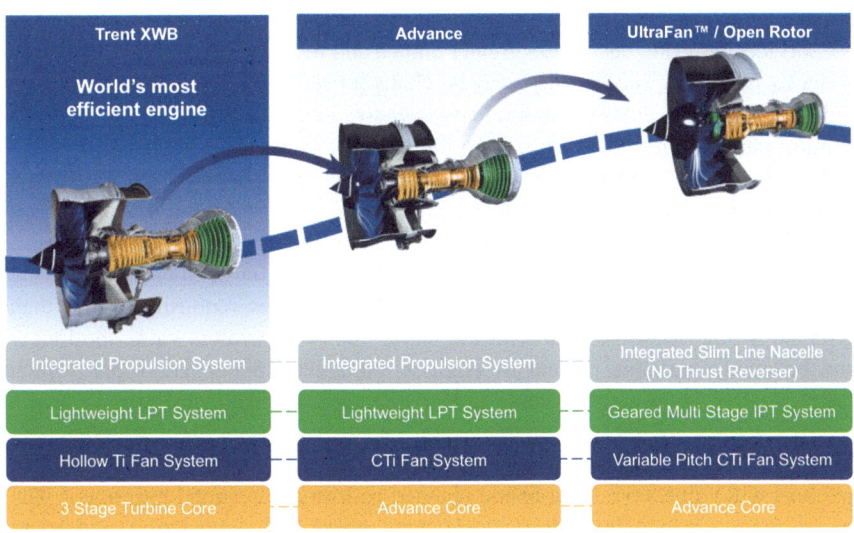

Fig. 6 Product evolution. Source: Rolls-Royce plc Marketing material, 2016

Enabling technologies for these engines are being developed at pace—a demonstrator 'Advance' engine will run in 2016, while the world's largest 3D printed aero-engine component is currently being tested as a prototype on the Trent XWB-97 flying test-bed—an Airbus A380 aircraft with the Trent XWB-97 engine installed in place of one of the aircraft's standard Trent 900 power plants.

Technology applied to aero-engines must provide value to airlines, so reduced fuel burn and improved environmental characteristics such as noise and emissions are critical. Through the history of airliner gas turbines, innovation has focused on increasing propulsive efficiency through higher bypass ratios, and improved thermal efficiency in the engine 'core' through ever higher temperatures and pressures. The first of these metrics drives engine design to larger fan sizes, requiring lightweight, high strength fan designs such as the new CTi fan design now completing development. Increased core temperature and pressure drive demand for ever more advanced materials in the engine's hot section, and the introduction of more effective turbine blade cooling systems.

Services innovation will be just as important, ensuring these engines provide reliable and cost-effective power for airlines—but this innovation also applies to current engine fleets, while product innovation tends to be more effective on new engine designs where physical constraints such as fan size can be allowed for in a matching new aircraft design.

The in-service Trent and RB211 fleets are maturing, and drive differing customer requirements across the product lifecycle. Today, the Trent fleet has an average age of just 7 years, but Rolls-Royce will be increasingly expected to deliver new services, offering customers greater choice and flexibility across all stages of a product's lifecycle. These stages can be characterised as a 'growth' phase, as an engine type starts to enter the market and fleet size increases rapidly, followed by a 'mature' phase where new engine entering the fleet are balanced by older engines retiring, and then a final phase once production has stopped, when the fleet gradually reduces in size as aircraft reach the end of their operating lives.

In the first of these three phases, operators typically value the risk transfer offered by TotalCare: they pay a constant services rate, charged per engine flying hour, while Rolls-Royce accepts the risk of any engines requiring earlier than predicted or high-cost overhauls. In the mature phase, operators value the longer time between overhauls typical of engines maintained under TotalCare, which benefit from the expertise available to Rolls-Royce as a result of extensive health monitoring data combined with the engineering expertise within the company. Trent engines are only now starting to enter the final phase of their lives, where operators will value a predictable and planned pathway to eventual retirement. The new services planned to address these three phases are now being offered to airlines.

In 2015, Rolls-Royce announced the first 'TotalCare® Flex®' agreements with AerCap and South African Airlines, Cathay Pacific and BMI Regional. TotalCare Flex offers customers services for engines approaching their retirement, with all the benefits customers have come to expect of TotalCare. Under this program, engines will be managed to the end of their operating life in a planned manner, designing engine support and shop visits to only provide the life on wing required to reach the end of the engine's operational life. This contrasts with TotalCare Life, the service product for aircraft operating in earlier phases of their life, where each engine shop visit is designed to achieve the longest possible time on wing, which will produce the most efficient outcome for operators flying aircraft over long periods.

The next innovation is the launch of SelectCare™. This event-based service enables customers to pay an agreed price for an engine overhaul when the overhaul occurs; in addition, the customer can select from a range of service options and pay for these on a $/engine flying hour basis. In January 2016, this new service was launched with American Airlines. SelectCare still delivers risk transfer on shop visit costs, while the operator accepts the risks of any unpredicted shop visit timing.

For airlines who wish to manage their own engine fleet, simply buying engine overhauls when required using a time-and-material approach, Rolls-Royce offers 'MRO Services'.

All these services use an extensive and expert network of overhaul bases: the majority of these are operated as stand-alone joint-venture companies, with key industry leaders as partners. Such joint-venture overhaul bases today include N3 in Germany, SAESL in Singapore and HAESL in Hong Kong.

When Rolls-Royce pioneered TotalCare in the 1990s, the focus was primarily on fixing components or engines when they became unserviceable. Today, focus is shifting to aircraft availability: the goal is to make sure customers' aircraft can depart from the gate and arrive at their destination on time at least 90 % of the time (compared with just 45 % underlying on time performance today).

This represents a step change shift to the way customers' operations are supported. Digital capabilities are key as Rolls-Royce moves forward with new services. "Big Data" is not a new concept—it is now widely referred to as part of the data revolution. Today, Rolls-Royce already analyses billions of data points from flights all over the world, and similarly Airbus is able to collect and analyse key data from the aircraft. In a world where the airline customer is looking for ever greater availability of their aircraft, the opportunity is clear for the aircraft manufacturer and engine OEM to continue to develop a collaborative approach to maximise value for their mutual customers.

Reference

All material in this chapter has been produced using Rolls-Royce plc Civil Aerospace's Marketing departments databases, analyses and explanatory material.

Peter Johnston joined Rolls-Royce in 1990, as an aircraft performance analyst, after working in aerodynamics and marketing with British Aerospace (now Airbus UK).

He has since worked in various marketing and engineering roles in Rolls-Royce and its' former subsidiary companies, including BMW Rolls-Royce, where he was heavily involved with launch of the Boeing 717, and International Aero Engines, where he was responsible for Product Strategy.

From 2004, Peter took responsibility for Business Development for power plants on a variety of new aircraft programs. In 2010, he moved to the Marketing team, and was then appointed Head of Customer Marketing, Airbus Programs. In this role, Peter led Marketing on all Trent 700, Trent 500 and Trent 900 opportunities.

From 2015, he has led Rolls-Royce's Marketing on all Airbus programs as Vice President, Marketing. In addition, he is responsible for Rolls-Royce's customer-facing marketing activities with airlines.

Peter holds a B.Sc. (Hons) degree in Aeronautical Engineering Science from the University of Salford, is a Member of the Royal Aeronautical Society, and a Chartered Engineer.

The Material Value Chain Services in Commercial Aviation

Jörg Rissiek and Mikkel Bardram

Abstract Material services are of fundamental importance in commercial aviation. While airlines acquire an aircraft as an input factor for their network operations only once, they will need spare parts over the full operational life cycle of an aircraft for decades—to keep it flying as seamlessly as possible with a minimum of interruptions caused by missing parts. Accordingly, the availability of the right spare part—to immediately replace the defective part at the right place—at the airport or maintenance base where the aircraft is waiting to go back into operation is crucial. Material services companies have the ambition to serve both material suppliers and airline customers with an efficient value adding global network and integrated performance based services, efficiently connecting both ends of the market. In order to reduce the significant transaction costs in this highly complex value chain, innovative airlines, as well as Maintenance, Repair and Overhaul (MRO) organisations, seek integrated material solutions, where a strategic partner organizes the full material value chain towards suppliers and provides a comprehensive part number range to the airline customers. In a consolidating market, leading material integrators—like Satair Group—will continue to grow and integrate the customer platform, the supplier interface, the value chain processes, and the digitalization-driven data management systems realizing functional and global economies of scale, scope, and density in combination with significant transaction cost reductions. New market segments such as used parts and additive manufacturing/3D printing will contribute to more opportunities along the material value chain. The strategies of the market players in combination with the market and innovation trends will reinforce each other, leading to a high integrative market dynamic for the years to come.

J. Rissiek (✉) • M. Bardram
Satair Group, Copenhagen, Denmark
e-mail: JOR@satair.com; mib@satair.com

1 The Structure of the Aviation Material Services Market

The commercial aviation material services market is of fundamental importance to airline customers. While the airlines acquire an aircraft as an input factor for their operations only once, they will need spare parts over the full life cycle of an aircraft up to 40 years—to keep it flying as seamlessly as possible with a minimum of interruptions due to missing parts. We will look into the current market structure, present the business models that exist in the market and give our view of the key trends driving this market in the long term.

With a modern long-range passenger aircraft containing some three million different parts, and spares manufactured by thousands of different suppliers, it is not surprising that the global value chain has become hugely complex. Moreover, compared to other industries such as automotive, the stakes are much higher. The consequences of deficiencies in the material supply are much more dramatic in civil commercial aviation because of the mere fact that a car can be more easily replaced than an Airbus A380 aircraft with 600 passengers. Therefore the players in the commercial aircraft services industry have to be even more reliable and much more active avoiding to leave an aircraft on the ground.

Unlike a rental car from *Hertz* or a shared car from *DriveNow* that you can leave on the roadside after a material failure and simply provide/take the next available one nearby, a grounded aircraft is a high value asset that can be only replaced at significant cost. The top priority target is for it to fly again immediately. Grounding an aircraft for just one day can cost an airline up to 600,000 US Dollars for a scheduled intercontinental flight. Passengers miss connections and need to get hotel rooms, time critical cargo deliveries are delayed, so compensation has to be provided. In some cases, a replacement aircraft has to be flown in because of non-availability of an essential no-go-spare-part. The airline reputation suffers with each significant flight delay and each cancellation (Aero Time 2015).

In order to understand the structure, complexity and size of the aerospace material services market, we will introduce the different actors in the material market and describe the material flows. Based on this understanding, we will explore the objectives and the market behaviour of the different actors and their respective strategies. In the civil aviation material market, a complex global material value chain links the different aircraft material category suppliers with the airline demand. Figure 1 illustrates this market and its key actors, showing the material flow from material suppliers to airlines with their material demand per aircraft or per aircraft fleet.

The total aircraft material market size as illustrated in Fig. 1 is about 42 billion US Dollars in 2016. Over the next 10 years, there will be around 4% estimated growth per annum reaching 62 billion US Dollars by 2025 (Airbus 2016; Broderick 2015; Shay 2015). The market figures are based on the Maintenance, Repair, and Overhaul (MRO) market analysis and forecast figures by Doan (2015), Marcontell (2016), Stewart (2015) and Berger (2016), as well as our own analysis. The market includes MRO for very large aircraft, wide body aircraft, standard body aircraft,

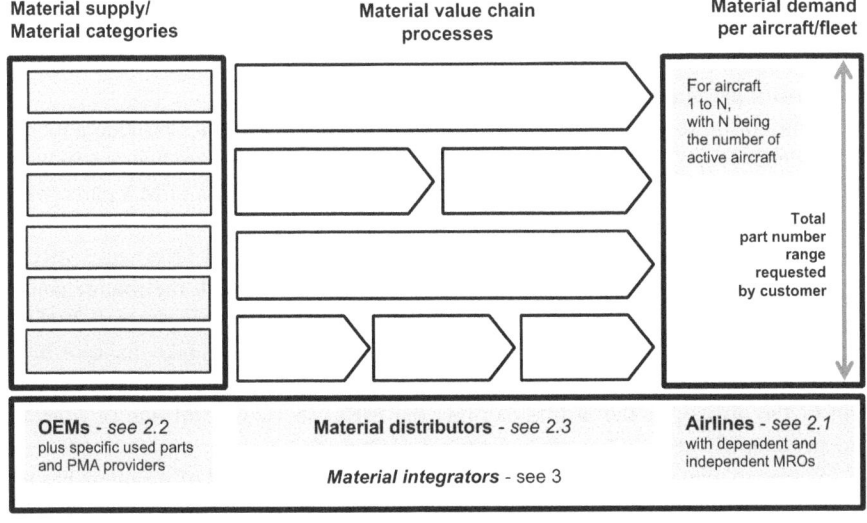

Fig. 1 Material services market structure. Source: own chart by the authors (2016)

regional aircraft and turboprop aircraft in commercial aviation, with an estimated material share of 62 % and a labor share of 38 % of the total market size. The growth rates vary depending on the material category and the world region; the overall market development is driven by the global aircraft fleet (Bernardini et al. 2015; Airline Business 2015; Seidenmann and Spanovich 2015).

The main supplier parts categories are engine and engine-related parts, high value repairable aircraft components, airframe parts provided by the airframe manufacturers, expendable parts, consumable parts, aircraft-related tools and support equipment for maintenance. Parts are regularly provided by Original Equipment Manufacturers (OEM), which have developed, certified, and produced the aircraft parts, for example, engine parts by *General Electric, Rolls Royce* and *United Technologies/Pratt & Whitney*, component units by *Honeywell, Rockwell Collins* and *Thales*, and expendable parts by *Hartwell, Pall* and *Telair*.

Specific competitive supplier market sectors are used parts and replacement parts. As an alternative to the supply of new OEM parts, spare material can be provided as certified used parts. In the higher value material categories, there is a supply of used parts in addition to new spare parts offered by OEMs (Pozzi 2015b). The used parts market segment has significant potential; in 2015, the global market size was over 4 billion US Dollars with a forecast of over 6 % growth per annum projected in the mid-term.

Furthermore, specialized companies with approval by the aviation authorities for replacing OEM parts by reverse-engineered, so-called Parts Manufacturer Approval (PMA) parts are offering material categories in competition to the OEMs (Honeywell Aerospace 2015). They range from high-value engine parts to simpler aircraft cabin parts. These companies are producing specifically PMA spare parts and are offering them in parallel to the OEM parts (Horwitz 2015; Chong 2015). The market size is less than 0.6 billion US Dollars in 2015 with limited under 2 % growth per annum projected in the mid-term.

In both of these specific market segments, dedicated companies provide a niche part number scope directly to the market. From our perspective, used parts will be a much stronger driver of the market structure and performance than PMA parts over the next decade.

The airlines always use approved Maintenance, Repair and Overhaul (MRO) organizations to place spare parts into their aircraft in line with the maintenance programs during the scheduled or unscheduled maintenance events (Russell 2015b; Delmas 2015). Depending on their organizational model, airlines take make-or-buy decisions on the material demand: either the MRO can be an internal organizational unit of the airline, or the airline can buy the MRO services from one or several independent MROs as a contractual partner (James 2016).

In order to manage the different material categories, a myriad of suppliers has to meet the global demand for all aircraft types—distributors like Satair Group provide material value chain processes as shown in Chart 1. These include the information and logistics flow between the market players, the planning of the material demand, purchasing, ordering, storage, logistics and quality management processes. These processes are provided for all fast- and slow-moving spare parts over a forward-looking time horizon in order to increase the efficiency of the material value chain, thereby connecting both ends of the material services market from Fig. 1, the material manufacturers as suppliers and the airlines/MROs as customers.

2 Strategies in the Material Services Value Chain

2.1 Customer Demand for Material Services

The objective of airlines as final customers is to keep each of their aircraft flying without any operational interruptions due to missing parts. Accordingly, the availability of the right spare part—to immediately replace the defective part at the right place—at the airport or MRO base where the aircraft is waiting to get ready for the next take off or is in a scheduled maintenance event is crucial for the airline. At the same time, the airlines and their MROs target reducing maintenance costs. Scaling this cost factor back means to trade off parts availability versus value chain costs.

When significant inventory investments are made, more parts will be directly available for maintenance events and aircraft availability will be high. Nevertheless, the inventory planning for material demand is highly challenging due to the several hundred thousand different parts required, depending on the type and the configuration of a modern jet aircraft. Infrequent parts usage for most of these parts over the aircraft life cycle is the driving factor for the availability and cost of spare parts supply: only around 10 % of parts are replaced regularly in an aircraft and for that reason move with some predictable frequency for an aircraft fleet. These parts can be forecasted and planned accordingly using statistical analytical tools, while the long remaining tail of very slow moving parts are more difficult to predict and the likelihood that they will never need to be replaced is substantial. However, if such a statistically very slow-moving part (because of their main characteristic, the Airbus terminology for such parts is Long-Term Storage parts) fails or is damaged, the operator expects—and the manufacturer has the contractual obligation to deliver—a replacement part. Therefore, different global value chain strategies for fast-versus slow-moving parts have to be applied.

Responding to this trade-off challenge of material availability versus material cost, the airlines/MROs adopt different purchasing approaches, which reflect different value chain strategies. These strategies differ across customer types and their make-or-buy approaches, but they can also be different for the respective material categories or the materials purchased for a specific aircraft fleet. One large airline, for example, is following a traditional in-house purchasing approach for an established legacy single aisle fleet, while at the same time it has outsourced a significant part of the material supply including the related services for a new state-of-the-art aircraft fleet.

Purchasing patterns can be structured according to the degree that the airline/MRO has to order each part that is needed, get the information on this specific part, and negotiate the price individually. They can opt for contractual agreements covering major factors such as lead-time, performance level, price, related services, conditions, incentive schemes, in a more complex contractual setting with a tactical or even strategic partner over a longer period. Depending on the level of particular versus integrated purchasing approaches, three typical strategies can be recognized in the market.

Airlines have traditionally used an ad hoc part-by-part purchasing approach within their organisation. This means an operational buyer gets information on a specific demand from the maintenance organisation that an order is required because of an actual parts failure or for a time-limited replacement part, and that the part is not stocked at the customer's maintenance base or not available at a remote flight destination's line maintenance station. The airline and its MRO might have a high exclusive spare parts inventory, but of course not a full stock of all potential spares, including slow moving parts with unpredictable demand. This would not be operationally and economically efficient for airlines because of the prohibitive cost of a complete stock with the low probability of a specific demand for the airlines' fleet of most parts being slow movers. For scheduled and unscheduled aircraft maintenance, airlines use the available parts in stock or they buy the parts they do not stock ad hoc based on the parts availability, lead-time, and look for competitive parts on the market, checking prices and conditions.

The costs for extensive exclusive stock keeping are high, and the lead-time for ad hoc purchases often causes delays for the maintenance period. As a result, operational requirements and cost pressure regularly lead to different approaches than pure ad hoc buying.

To reduce material cost and to avoid delayed material delivery, airline and MRO customers increasingly request specific performance-based contracts for a certain critical part number range or material category of their aircraft fleet. Such *Performance Based Services (PBS)* arrangements guarantee material availability and timely performance; the associated costs are linked to the agreed performance level, and are often combined with a penalty and incentive regime. The services are modular, tailored to customer specifications, and can vary in part number scope and service level ambition (This is the Airbus Group definition of Performance Based Services, see Matrat et al. (2014)). PBS contracts imply an agreed price, confirmed lead-time and availability for the contractual part numbers. The customer can increase the stability of the maintenance schedules due to guaranteed parts availability for the negotiated spare parts at their main hubs as well as their outstations. Airlines are outsourcing material packages and the related value chain via PBS to MROs and distributors.

Specific dedicated performance contracts, and even more ad hoc orders for each missing part, require a high organisational effort to get the best prices and conditions. It involves negotiating the contract scope and terms, monitor delivery performance and potential contractual incentive schemes, as well as adopting ad hoc purchasing and contracts to a changing fleet network over time.

In order to reduce significant transaction cost in the material demand, several innovative airlines seek a fully integrated material solution. A strategic partner organises the full material value chain towards all suppliers and the full part number range needed to operate the customer fleet. In this case, the airline relies on one integrator to manage all materials and to deliver all spare parts as needed for the aircraft fleet considering all dimensions of the total cost of ownership along the value chain. As a general trend, customer purchase behaviours are changing from ad hoc to long term outsourcing arrangements with an increasing part number scope covered (Stewart 2015; McFadden and Worrels 2012).

2.2 Supplier Approaches to the Material Services Market

On one hand, material suppliers produce their material quantity to be built into aircraft on final assembly lines of aircraft manufacturers for the initial entry into service, and on the other hand, for the related spare parts demand over the operational life cycle of an aircraft. A supplier/OEM wants to ensure the services market penetration of his products. He wants to keep or even increase his market share over the life cycle, by securing a high availability of his parts for customers, and the resulting customer satisfaction helps building specific OEM brand name capital and long-term business success.

The supplier can go for different services strategies. There are different ways to achieve the objective to be able to supply parts throughout the whole product life cycle. Each supplier/OEM has to decide how to configure their specific approach by carefully evaluating each of the activities in the material value chain. They do not have the option to neglect the services market; this would lead to massive customer interventions and damage to their reputation, with a negative impact on new aircraft production sales and for the OEM company revenues.

The supplier can employ a pure *make-approach*. He can build his own material network and distribute his full material scope in a dedicated global service organisation. Several major suppliers have historically built such an in-company distribution network for their products to cover the spare parts demand worldwide.

While looking for the best way to sell and promote their products during the life cycle, suppliers will regularly consider going on their own into the services market. By promoting and selling directly to their global customer base, they control direct access to and communication with customers and get immediate, direct performance feedback. Making this choice, they have to consider the high investment and permanent operational cost of such a dedicated distribution network limited to their own parts.

The material supplier can also decide to go into the services market with a *partner-approach*. He can leave the major services activities to such a partner, or to several partners in specific world regions, or handle dedicated parts packages differently. A significant supplier has granted Satair Group, for example, the distribution of his parts in one world region, while keeping distribution insourced in another region. Satair Group applies different distribution and cooperation agreement structures to match the specific supplier's strategy. Many are globally exclusive, while some are exclusive for a certain region.

The supplier can decide to give the full services business to a trusted distribution partner, or he can go for a hybrid solution with parallel distribution channels. A specialised high technology supplier will also consider focusing on his core innovation, development and production strengths, and will outsource his services distribution to a business partner. This business partner can use the full sales force to promote the spare parts to the airline; this includes established global distribution, full service logistics and a marketing network, and the partner can help the supplier to be more successful in the long term by providing his services expertise and network.

Small to mid-size OEM companies might have limited investment resources and have to select the best use of their capital. These OEMs might decide not to use their investment opportunities and their specifically qualified human capital to build a stand-alone sales/distribution network. Large companies might want to take a strategic approach to focus their investment resources on their core activity of development, engineering and production capabilities, to expand and diversify their product range, and/or to develop new innovative products for in-production as well as under-development aircraft.

Currently, several suppliers are reassessing their strategic material distribution approaches. Some large players tend to insource distribution to get direct control of their own service offerings and keep a market share for their own products over time. Others outsource in order to improve the services performance over the operational life cycle and to focus on their current and future core product development, engineering and production activities outside the material distribution services market. If suppliers decide to outsource, they rely strategically on distributors as their value chain services partners.

2.3 The Role of Distributors in the Material Value Chain

Distributors are the intermediaries covering the material value chain processes to connect suppliers and customers in the most efficient way by using economies of scale and scope (Shepherd 1997; Schmidt 2012). Their business model is focused on material services markets. Distributors typically target to add value to both OEMs and the airlines covering a wide business portfolio. Distributors help OEMs drive revenues, reduce costs and improve cash flow. Representing many OEMs on a global scale leads to higher operational efficiency and less transaction cost. We see transaction cost savings as major driver for efficiency increases by distributors/integrators (Williamson 1979; Williamson 1985; Sedatole et al. 2011). A distributor has a specialised sales force and a dedicated value chain network designed for a wide parts portfolio, not only for one limited supplier parts range. A distributor can therefore share his structural costs for global infrastructure and specific human capital over a large parts portfolio and shall be more efficient in his global set-up.

What are the customer benefits of working closely together with a specialised distributor? A distributor tailors the material services to customer requirements in terms of regional availability and performance level. He can use global resources to provide dedicated as well as modular services adapted to specific distribution demands. The customer has less capital binding investments in his own parts inventories and related infrastructure, resulting in financial relief for him, as his balance sheet is less burdened. He reduces his risk of over-stocking or incorrect stocking. If he wanted a comprehensive safety stock, he would need to speculate on stocking slow movers for his limited fleet.

The customer gains control over the material costs through a performance contract, ideally with a steady, predictable cost flow over the full aircraft life cycle. The distributor becomes a risk-sharing partner, but one who can spread the inventory and planning risk over many customers and their fleets. The specific distributor experience and material expertise is immediately available and the customer can participate in learning curve effects over time. The distributor's material-specific experts, processes, assets and infrastructure are fully available for an efficient supply to the airlines.

In parallel, distributors reduce transaction cost for the market players on both the supply and the demand side, by their expertise

- in getting specific information on the services regional markets and their conditions,
- negotiating in a specific cultural and technical environment,

- performing in a complex supply chain system, and
- keeping the performance in a dynamic environment stable, while
- developing contractual relational agreements and operational links
- building on a long-term collaborative, trustful relationship.

Distributors can systematically use the received feedback on product performance and further improve the operational value chain network and part availability. For the customers, the distributors can drive down value chain costs by offering a wide product portfolio with a short lead-time. This helps to lower both transaction and inventory costs for the airlines.

Satair Group, as an industry-leading material management services provider, has a dedicated sales and product management organisation to build and develop a trust-based and open-book relationship with their suppliers/business partners. Satair Group has an in-depth understanding of customer demand for the specific material categories per aircraft fleet and per world region through its global value chain network. It provides accurate forecasting data based upon its wide experience in inventory planning and optimization.

With its firmly established global footprint, services centre network in place covering all world regions, Satair Group is realizing significant economies of density: providing culture-specific customer proximity, connected warehouses, strong data management, logistics, planning tools for a wide range of parts from many OEMs in each region. A recent example is the new, operational Singapore Seletar Services Centre covering the Asia-Pacific region. Figure 2 shows the warehouse of the Singapore Centre.

With this connected value chain set-up, Satair Group generates scale in its global operating model; the structural costs are split across many OEM product lines and for hundreds of airline customers. With its global supply chain set up, Satair Group is responding to infrequent, fragmented individual parts demand, to the lack of data and inventory transparency across the regions and players, using planning models of inventories to buffer the market inefficiencies. An integrative market player like Satair Group bridges the multiple challenges of the material value chain and its core strategic purpose has made the market more efficient for the suppliers as well as the customers (Rinn 2015).

3 Future Trends in the Material Market

The services market is currently maturing and developing very quickly due to several new market and innovation trends in the industry affecting the business model and the structure of the aviation services material flow (Spafford et al. 2015; EADS 2013; Rissiek and Kressel 2004). The main structural trends influencing the material market are:

- Increased *outsourcing* from both customers and suppliers leading to *integrated offers* and *consolidation* in the market structure

Fig. 2 Warehouse of Satair Group Services Centre in Singapore. Source: Picture by Airbus/Satair Group

- Increasing market penetration of *used parts*
- Integration and network effects of *digitalization*
- New value chain set-ups via *additive manufacturing*

Airline customers are increasingly *outsourcing* their material management to qualified partners, in order to realize efficiency gains along the global value chain, to save operational costs and to rely on the expertise of a dedicated specialist in value chain management. The main drivers for this outsourcing trend for airlines are:

- Airlines want to focus on their core tasks: passenger and cargo transportation,
- Airline budget restrictions and cost pressure endanger affordability of classical ad hoc procurement and operations,
- Desire to reduce financial exposure by limited capital commitments for inventory,
- Need for enhanced expertise and capabilities in a complex global support set-up,
- Access to leading innovative technology/solutions/networks,
- Strong pressure to ensure highest operational aircraft availability by having an extremely high material availability,
- Especially for critical unscheduled maintenance events, not planned in the regular aircraft maintenance schedule.
- Flexible value chain designs considering specific airline networks and regional requirements with variability of the cost base.

The outsourcing trend can be seen in a similar way on the supply side: Suppliers want to focus on their core research, development and production business of often high-technology aircraft parts, they are looking for less capital exposure compared to an in-house distribution organisation to secure their parts distribution services in the market.

A distributor such as Satair Group helps to manage the challenges resulting from such an outsourcing strategy: long-term performance and services commitments require flexible, long-term contracts, high-risk control, and trustful customer relationship with open information exchange. In the mid- to long-term, requirement adaptations, material operation with variability of the cost base are implemented in a collaborative way, as the airline develops new routes/regional markets, acquires new aircraft, shifts its continental/intercontinental network, and retires entire aircraft types/fleets.

This trend will lead to *integrated offers* covering a very wide range of parts in all material categories as airline and MRO customers are increasingly looking for integrators to ensure transaction cost efficiency. Some customers will fully give all material processes to one integrator.

Integrated customer offerings for Airbus proprietary parts started in Airbus/Satair Group with the *Airbus Managed Inventory (AMI)* program in 2010. AMI consists of an automated inventory replenishment solution, which helps Airbus customers reduce their inventory holding costs. By capturing material consumption information in real-time and automatically triggering replenishment orders within the agreed inventory levels, the service guarantees high on-shelf part availability, while decreasing the overall inventory stock level, and reducing the cost of ordering Airbus material (Aviation Pros 2015).

Satair Group started a strategic initiative to increase significantly its integrated offering beyond focused AMI. Besides a modular contractual part number offering—*Integrated Purchasing Program (IPP)*, as well as an offering for high-value repairable components *Airbus Flight Hour Services (FHS)*, the focus is now on the growth of *Integrated Material Services (IMS)* as the comprehensive material offering of Satair Group. This integrative, performance-based approach has the following characteristics to respond to market trends:

- IMS covers all part number demand as requested by the customers, currently for a range from 4000 to 40,000 part numbers,
- in a customized, modular approach considering the specific factors influencing the demand of each customer,
- while not all customers are opting for the fully integrated mode now, there is a clear trend towards a more comprehensive range covering several material categories,
- and IMS provides a competitive advantage as the offer reduces the total cost of ownership at a guaranteed performance level significantly, plus
- IMS also includes additional modular services like parts repair, maintenance tools, as well as customized modification kits.

While offering more and more integrated services for all material categories, the business model will change for distributors following the above trends. They are becoming *material integrators*. While the leading distributors are becoming bigger and improving their offerings along the value chain, the market will consolidate to meet the integrators' expectations.

Market structure consolidation in a still fragmented material market—with many limited scope specialists and only local focus—will take place over the next decade. To drive down the value chain costs, significantly realizing economies of scale and scope in combination with transaction cost reduction, further consolidation will need to happen. Boeing already acquired the distributor Aviall in autumn 2006 (Moorman 2008; Raley 2015; Ostrower 2016). Recent acquisitions in the distributor market have been the acquisition of distributors like *Interturbine* by *B/E Aerospace/KLX Solutions* and of *Haas* by *Wesco Aircraft Holdings* (Canaday 2013; Russell 2015a).

Airbus acquired *Satair A/S* in 2011 to cover a wider material category scope and become market leader in an integrator mode, while combining the Airbus material and logistics organization with the Satair distribution organization in the newly founded Satair Group (Flightglobal 2011; Satair Group 2015b). In 2015, Satair Group integrated *Eltra Aeronautics*, a specialized mid-size spare parts distribution company with its main base and logistics facilities in Singapore, into its material services network. With Eltra Aeronautics having a dedicated regional market reach in Asia-Pacific, including China, Satair Group aims to gain even more momentum in this rapidly growing region.

The *Used parts business* is already an important market category and will continue to grow steadily over the next years (Pozzi 2015b; Satair Group 2015a; Honeywell 2015). The market is seeing an increasing number of aircraft retirements based on the airlines' and leasing companies' investments to replace legacy aircraft by more modern and fuel efficient fleets of up to 3 % per annum of the flying global fleet in the mid-term. Only in the short term the retirement trend might be slowing down to a certain extent, as some airlines and leasing companies are keeping older aircraft for a limited time due to a current macroeconomic environment with very low oil and fuel prices, while low interest rates make fleet roll-overs cheaper. This roll-over/retirement trend will lead to a much higher market presence for more and for a wider range of used parts coming out of dismantled aircraft.

This trend will make used parts a more important, specific material category with dedicated value chain elements, especially through dismantling and re-certification processes for additional used parts coming from retired aircraft. An integrators' used parts value chain will offer used parts as complementary material category in its portfolio and will optimize its portfolio considering the growing role of used parts and their specifities.

Digital technologies are emerging and are applied more systematically: this implies increasing data collection by sensors for performance monitoring systems, overseeing potential aircraft and sub-systems parts failures over the aircraft's operational lifecycle. They are based on fully connected digital devices and advanced analytics systems. When integrated with real-time information for the ongoing processes, digital data analytics will create a new dimension of optimization. Data accessibility and usability along the value chain will result in an automated, predictive approach for parts deliveries and active inventory planning

(Gubisch 2015; James 2015; Coutts 2014). The network of airline fleets, parts suppliers, MROs and the material integrators' processes will be more strongly connected, even in real time. The increased digital transactional traffic between the actors will again create data that generate deeper insights valuable for customers and suppliers, as well as the integrators will be able to improve their global material network based on the digital analytics dynamic results.

Digitally enabled platforms will generate additional economies of scope, and economies of density for those integrators, who can derive added value via digital efficiencies in the value chain or provide additional digital-based services and network-based business models. Predictive maintenance and smarter inventory planning systems based on data analytics will further link the material value chain with the customer demand, also in an anticipating predictive way. Such integrative ecosystems will further boost integration efficiency along the value chain because the integrators will have a new dimension of data and forecasting tools at hand to connect to both market sides, suppliers and customers. At the same time the amount of valuable digital information the integrator provides as structured feedback for improvements to the suppliers and customers is exponentially increasing—a digital virtuous circle is created (Pozzi 2015a; General Electric/Accenture 2014).

Additive manufacturing has strong innovation potential to change the set-up of material production and availability for thousands of part numbers for airframes, engines, cabins and equipment. Additive manufacturing is a technology that has recently shown its potential to revolutionize entire industries, it will have strong effects on the material value chain of aircraft, too (D'Aveni 2015; Armstrong 2015; Pierobon 2016). The term additive manufacturing (used here synonymously with the term 3D printing) groups a range of different technologies and materials, where parts are built layer-by-layer based on a digital model. The additive manufacturing principle of printing layer-by-layer is maintained with all processes for both plastic and metal printing: either material is dispensed and added through a nozzle, or material powder is melted by a laser or an electron beam. In comparison to additive manufacturing, traditional subtractive manufacturing cuts material away from a solid piece of material, for example, by metal machining, to create an object.

The benefits of additive manufacturing in the commercial and military aviation industry include faster lead times, especially for out-of-production or individual parts, less complex and lighter designs, avoiding expensive production tooling (Rissiek 2016). The main driver for using this technology specifically in the services market is the flexible and efficient production of smaller volumes, as well as the opportunity to offer customized solutions to airlines, for example, a recently additive manufactured part for a new *Qantas* cabin solution (Rehmanjan 2015). First additive layer manufactured parts were applied to in-service aircraft in 2014; Airbus and Satair Group are continuously expanding the application scope, including new customized services. While additive manufacturing might even enable new logistics and business models, such as local on-site production, it is still in an early certification and rollout phase. Additive manufacturing will show its full potential as a new stream in the value chain serving specific customer demands in the next years. Consequentially, additive manufacturing will reduce storage for an increasing number of parts.

With the ambition to serve both material suppliers and customers with an efficient value adding global network and integrative performance-targeting processes, it is vital to connect both ends of the market efficiently in a highly dynamic environment. Leading material integrators—like Satair Group—will continue to grow and integrate customer platform, supplier interface, value chain processes and increasingly digital data management systems realizing functional and global economies of scale, scope, and density in combination with significant transaction cost reductions.

New market segments, especially used parts and additive manufacturing, will both add more complexity and contribute to opportunities to further optimize the material value chain networks for commercial aircraft fleet. The market players' strategies and the future trends outlined in this article will reinforce each other and lead to a much higher integrative market dynamic in value chain services for the years to come, compared to what we have observed to this point.

References

Aero Time (2015) Human error—supply chain's worst enemy. Available via http://www.aerotime.aero/en/mro/mro-news/other/16488-managing-the-mro-supply-chain. Accessed 22 Jun 2016

Airbus (2016) Global Market Forecast (GMF) 2016–2035—Mapping demand. Available via http://www.airbus.com/company/market/global-market-forecast-2016-2035/

Airline Business (2015) Special report aircraft & engines 2015. Airline Business, Apr 2015. Available via http://s3-eu-west-1.amazonaws.com/fg-reports-live/pdf/Aircraft%20&%20Engines%202015.pdf. Accessed 22 Jun 2016

Armstrong D (2015) Layering up. In: Russell S-J (ed) MRO Yearbook 2016. London, pp 48–51

Aviation Pros (2015) Thai airways chooses to implement Airbus Managed Inventory (AMI) service. Available via http://www.aviationpros.com/press_release/12093947/thai-airways-chooses-to-implement-the-airbus-managed-inventory-ami-service. Accessed 22 Jun 2016

Berger JM (2016) MRO market forecast and market trends. In: ICF International (ed) Presentation at the airline engineering & maintenance China & East Asia Conference, Singapore. Available via http://www.icfi.com/insights/presentations/aviation/mro-market-forecast-and-trends-asia-pacific. Accessed 22 Jun 2016

Bernardini E, Mauri M, Fabre P (2015) Gaining altitude—with help from strong tailwinds, global aerospace and defence industry outlook. In: Alix Partners (ed). Available via http://www.alixpartners.com/en/Publications/AllArticles/tabid/635/articleType/ArticleView/articleId/1801/Gaining-Altitude-With-Help-from-Strong-Tailwinds.aspx#sthash.qvO4uQ6K.dpbs. Accessed 22 Jun 2016

Broderick S (2015) Growing more concentrated. MRO forecast underscores increasing domination of fewer aircraft types, importance of end-of-life services. Aviation Week & Space Technology—MRO edition, Apr 2015, pp 8–12.

Canaday H (2013) Aftermarket parts market consolidation likely to increase. Available capital and broader services driving aftermarket consolidation. Aviation Week & Space Technology—MRO edition. Available via http://aviationweek.com/awin/aftermarket-parts-market-consolidation-likely-increase. Accessed 22 Jun 2016

Chong A (2015) Part works. Airline Business, Dec 2015, pp 26–29. Available via http://s3-eu-west-1.amazonaws.com/fg-reports-live/pdf/AB%20Issue%20Dec%202015.pdf. Accessed 22 Jun 2016

Coutts J (2014) Innovations: big analytics. Quand l'intelligence vient aux données. In: Thales (ed) Innovations. Spring 2014, pp 10–13

D'Aveni R (2015) The 3-D printing revolution. Harvard Business Review, May 2015, pp 40–48. Available via https://hbr.org/2015/05/the-3-d-printing-revolution. Accessed 22 Jun 2016

Delmas C (2015) Scheduled maintenance requirements. Maintenance programmes and planning. FAST Flight Airworthiness Support Technology—Airbus Technical Magazine, 55, pp 28–37

Doan C (2015) Turbulence ahead—Disengage autopilot 2015—2025 global fleet and MRO market forecast. In: Oliver Wyman (ed) Presentation at the MRO Europe Conference, London. Available via http://www.oliverwyman.com/content/dam/oliver-wyman/global/en/2015/oct/2015_2025_MRO_Forecast%20_Trends_MRO_Europe_Presentation_20151013.pdf. Accessed 22 Jun 2016

EADS (2013) Services by EADS—a vital business dimension. EADS, Leiden (Available via the author (jor@satair.com))

Flightglobal (2011) Airbus makes bid to buy Satair. Available via https://www.flightglobal.com/news/articles/airbus-makes-bid-to-buy-satair-359995. Accessed 22 Jun 2016

General Electric/Accenture (2014) Industrial internet insights report. Available via http://Assets/DotCom/Documents/Global/PDF/Dualpub_2/Accenture-Industrial-internet-Changing-Competitive-Landscape-Industries.pdf#zoom=50. Accessed 22 Jun 2016

Gubisch M (2015) Predicting change. Airline Business, Dec 2015, pp 22–24. Available via http://s3-eu-west-1.amazonaws.com/fg-reports-live/pdf/AB%20Issue%20Dec%202015.pdf. Accessed 22 Jun 2016

Honeywell Aerospace (2015) The pre-owned parts market: strong today, stronger tomorrow. Phoenix

Horwitz D (2015) Material benefits—a review of the PMA market. In: Russell S-J (ed) MRO Yearbook 2016. London, pp 40–43

James O (2015) Les big data ouvrent l'ère de l'aéronautique de service. L'Usine Digitale, Jun 2015. Available via http://www.usine-digitale.fr/article/les-big-data-ouvrent-l-ere-de-l-aeronautique-de-service.N334101. Accessed 22 Jun 2016

James O (2016) La maintenance sous haute tension. L'Usine Digitale, Feb 2016. Available via http://www.usinenouvelle.com/article/la-maintenance-sous-haute-tension.N376778. Accessed 22 Jun 2016

Marcontell DA (2016) Winds of change 2016–2026 global fleet and MRO market forecast. In: Oliver Wyman (ed) Presentation at the MRO Americas conference, Dallas. Available via http://www.oliverwyman.com/content/dam/oliver-wyman/global/en/2016/apr/20160405_OW_CVK_MRO_Americas_PresentationJS.pdf. Accessed 22 Jun 2016

Matrat B, Leukel A, Rissiek J (2014) Performance Based Services—PBS. Framework and definitions. Airbus Group, Blagnac (Available via the author (jor@satair.com))

McFadden M, Worrels DS (2012) Global outsourcing of aircraft maintenance. J Aviat Technol Eng 1(2):63–73

Moorman R W (2008) Aviall: stronger than ever. Air Transport World. Available via http://atwonline.com/operations/aviall-stronger-ever. Accessed 22 Jun 2016

Ostrower J (2016) Boeing ramps up push into the airplane parts business. Wall Street Journal. Available via http://www.wsj.com/articles/boeing-ramps-up-push-into-the-airplane-parts-business-1461539460. Accessed 22 Jun 2016

Pierobon M (2016) 3D printing—copy shop. MRO Manag 18:60–62

Pozzi J (2015a) A connected industry. Aircraft Technology Engineering & Maintenance, Jun/Jul 2015, pp 54–58

Pozzi J (2015b) All parted out. In: Russell S-J (ed) MRO yearbook 2016. London, pp 44–47

Raley D (2015) Accelerating growth. Boeing's strong commercial and defense businesses have an essential partner—services. Boeing Frontiers, Nov 2015, pp 18–30

Rehmanjan UH (2015) 3D printed parts for quick turnaround aircraft projects and legacy issues. Conference paper at the 15th aviation technology, integration, and operations conference, Dallas. Available via http://dx.doi.org/10.2514/6.2015-2739. Accessed 22 Jun 2016

Rinn T (2015) COO insights—service excellence. Roland Berger Publications. Available via http://www.rolandberger.com/media/publications/2015-12-03-rbsc-pub-COO_Insights_Service_Excellence.html. Accessed 22 Jun 2016

Rissiek J (2016) 3D Printing—an innovative technology shaping the future of aviation. Presentation at the Singapore Aerospace Technology and Engineering Conference (SATEC), Singapore (Available via the author (jor@satair.com))

Rissiek J, Kressel J (2004) New developments in purchasing and supply chain strategies for the aviation industry. In: Cooper E (ed) Business briefing: global purchasing & supply chain strategies. Business Briefings, London, pp 52–55

Russell S-J (2015a) Consolidation in action. In: MRO Network (ed). Published Apr 14, 2015. Available via http://www.mro-network.com/opinion/2015/04/consolidation-action/5196. Accessed 22 Jun 2016

Russell S-J (2015b) (ed.) MRO Yearbook 2016. London

Satair Group (2015a) Certified used parts keep aircraft flying. Simply Fly. Available via http://ipaper.ipapercms.dk/Satair/SatairGroupBrochures/SatairGroup/Magazines/SimplyFly3/. Accessed 22 Jun 2016

Satair Group (2015b) Excellence connected. Simply Fly. Available via http://ipaper.ipapercms.dk/Satair/SatairGroupBrochures/SatairGroup/Magazines/SimplyFly3/. Accessed 22 Jun 2016

Schmidt I (2012) Wettbewerbspolitik und Kartellrecht. Eine interdisziplinäre Einführung, 9th edn. R. Oldenbourg Verlag, Munich

Sedatole KL, Vrettos D, Widener SK (2011) Beyond transaction cost economics: the incremental effects of intra-firm moral hazard and management control mechanisms on strategic outsourcing decisions. Discussion paper. Michigan State University, East Lansing

Seidenmann P, Spanovich DJ (2015) Widebody MROs gearing up for next-gen airframes. Widebody MRO providers ramping up for A380, 787, 747-8 and A350 maintenance. Aviation Week & Space Technology. Available via http://aviationweek.com/mro/widebody-mros-gearing-next-gen-airframes?NL=AW-05&Issue=AW-05_20150106_AW-05_295&YM_RID='email'&YM_MID='mmid'&sfvc4enews=42&cl=article_4_b. Accessed 22 Jun 2016

Shay LA (2015) Expect more MRO activity in 2016 than 2015. Aviation Week & Space Technology—MRO edition. Available via http://aviationweek.com/mro/expect-more-mro-ma-activity-2016-2015. Accessed 22 Jun 2016

Shepherd WG (1997) The economics of industrial organisation, 4th edn. Prentice-Hall, Upper Saddle River, NJ

Spafford C, Hoyland T, Medland A (2015) MRO survey 2015—Turning the tide. A wave of new aviation technology will hit the MRO industry. Oliver Wyman Publications. Available via: http://www.oliverwyman.com/insights/publications/2015/apr/mro-survey-2015.html#.V2pqvVdyj80. Accessed 22 Jun 2016

Stewart D (2015) MRO market forecast & trends. In: ICF international. Presentation at the MRO Asia-Pacific conference, Singapore. Available via http://www.icfi.com/insights/presentations/2015/mro-market-forecast-trends-presentation. Accessed 22 Jun 2016

Williamson OE (1979) Transaction-cost economics: the governance of contractual relations. J Law Econ 22:233–261

Williamson OE (1985) The economic institutions of capitalism: firms, market, relational contracting. Macmillan, New York

Jörg Rissiek, Dr. rer. pol., is Vice President and Head of Corporate Strategy and Projects of Satair Group since July 2014. He is based at the company's headquarters in Copenhagen and is responsible for driving Satair Group's strategic approach and the execution of strategic initiatives to become the global market leader in the commercial aircraft material management business over the aircraft life cycle across aircraft platforms.

From 2011 to 2014, Jörg worked as Vice President in Corporate Development of the Airbus Group's headquarters in Paris/Toulouse, focusing on the Airbus Group's services strategy and development. From 2006 to 2011, he was the Vice President of Material and Logistics Operations in Airbus Customer Services at Hamburg Airport, managing the interface to 300 Aircraft operators. He joined Airbus after working eight years at Lufthansa, as Director VIP and Government Aircraft Customer Support and Director Purchasing Processes/Supply Chain Management at Lufthansa Technik. He negotiated major aircraft purchasing campaigns in Lufthansa's aircraft asset management organization.

Jörg has been a research affiliate in the Department of Economics at the Technical University Ilmenau, Germany. He has a degree in Economics, Diplom-Volkswirt, from Münster University with a specialization in Transport Economics.

Mikkel Bardram is the founding CEO of Satair Group. He is based at the company's headquarters in Copenhagen and manages Satair Group to become the global market leader in the commercial aircraft material management business over the aircraft life cycle across aircraft platforms.

In 2006, Mikkel joined Satair as Corporate Manager in Corporate Projects. In 2007, he became Vice President of Supply Chain with full responsibility for Satair's material value chain. From 2002 to 2005, Mikkel worked as an Engagement Manager with McKinsey & Co. Prior to that he had been SAP Specialist in Sales and Distribution at Novozymes.

Mikkel holds Master degrees in Science, in Economics and in Business Administration from Copenhagen Business School. He also holds a Bachelor degree in International Trade from Copenhagen Business School.

Predictive Maintenance: How Big Data Analysis Can Improve Maintenance

Jim Daily and Jeff Peterson

Abstract What if an aircraft part could tell you when it needs to be repaired or replaced? With continuous data collection, monitoring and application of advanced analytics, it can. Predictive maintenance in aviation offers the promise of increased reliability along with bolstered operational and supply chain efficiencies. The integration of Cloud-based analytics with industrial machinery has driven the advent of a new "Industrial Internet"—and with it, opportunities for huge productivity gains. Industrial machines equipped with a growing number of electronic sensors can see, hear and feel a lot more than ever before—and deliver enormous amounts of data to be analyzed. Sophisticated algorithms then provide insights that allow us to operate machines—fleets of engines, airplanes, and network entire systems like airlines and airports—in entirely new, more efficient ways. To realize the value in predictive maintenance and achieve business outcomes that matter, machines, data, insights and people need to be brought together.

1 Introduction

A generation ago, an aviation engineer would have been on the cutting edge of technology using small sets of data gathered after each flight to enhance the performance of equipment in service. This was the limitation of the technology at the time, and the inherent limitation in learning. Today, digital connectivity is removing that barrier.

The integration of cloud-based analytics ("big data") with industrial machinery ("big iron") has driven the advent of a new "Industrial Internet"—and with it, opportunities for huge productivity gains. Industrial machines equipped with a growing number of electronic sensors can see, hear and feel a lot more than ever before—and deliver enormous amounts of data to be analyzed. Sophisticated algorithms then provide insights that allow us to operate machines—fleets of

J. Daily · J. Peterson (✉)
GE Aviation, Evendale, Ohio, USA
e-mail: jeff.peterson@ge.com

engines, airplanes and entire network systems like airlines and airports—in entirely new, more efficient ways.

This is the Digital Industry; where business moves at a much faster pace, with the confidence and speed enabled by a new reality where physics meets analytics and all participants across the value chain win.

To unlock the value in predictive maintenance and achieve business outcomes that matter, you will need to bring together your machines, data, insights and people. To get started you will need to:

- Get Connected: connect machines, data and people for an integrated view of one's entire operations.
- Get Insights: use advanced data analytics to understand what drives factors such as overall equipment effectiveness, equipment waste, production quantity, inventory, and more.
- Get Optimized: optimize operations, maintenance planning and equipment reliability through predictive analytics.

2 From Preventative to Predictive

Preventive maintenance, a common industry practice, is maintenance that is regularly performed on a piece of equipment while it is still working, so that it does not break down unexpectedly and cause unplanned downtime. In aviation, this process is based on stringent requirements to repair or replace parts and systems before they fail to assure an aircraft's safe, continued operation. It relies on engineering data and operational experience to determine the appropriate point in time to make the repairs—such as replacing a part after 5000 flight hours, based on studies that indicate reliability cannot be assured after that point.

While preventive maintenance in aviation is largely responsible for the outstanding safety and reliability of today's aircraft fleet, its effectiveness is limited by the backward-looking nature of the analyses. It is based on history and statistics; it does not take into account the everyday operating conditions for a specific asset. Despite the fact that each engine, aircraft and system is subject to its own set of conditions and experience, a preventive maintenance schedule is simply based on averages. Due to the variability of actual experience, some good parts are replaced too soon, while others may fail before the prescribed replacement schedule.

In contrast, predictive maintenance uses advanced analytics based on the assets actual operating conditions. Monitoring for future failure allows maintenance to be planned before the failure occurs. Ideally, predictive maintenance allows the maintenance frequency to be as low as possible to prevent unplanned reactive maintenance, without incurring costs associated with doing too much preventative maintenance. It can also allow for a more optimized operation of a component.

Predictive maintenance eliminates the need to shut down for service on a periodic basis, enabling maintenance techs to monitor things like vibration, heat and fuel consumption to understand what's going on deep inside complex machinery. In fact, Fortune Magazine reports that PTC believes predictive maintenance will be one of the first killer applications of the Industrial Internet (Wollenhaupt 2016).

To enable predictive maintenance, today's aircraft are loaded with sensors, generating valuable streams of operational data. The key to predictive maintenance is applying analytics to those data streams to identify patterns and trends that can direct the maintenance strategies—delivering the right information at the right time in the right context to prevent failures. The data can also be mined for suggestions about future ways to improve product design.

From a maintenance point of view, the operational information coming from the sensors on a jet engine or aircraft would be fed into an analytics platform; over time, your data science team would be able to build sophisticated models and analytics that would predict when a machine is likely to fail. You would then initiate a repair before that point of failure. Eventually, this will drive towards a world of "zero unplanned downtime"—no more power outages, no more flight delays, and no more factory shutdowns.

As the industry becomes more comfortable with intelligently monitoring and analyzing equipment in order to determine the need for repair or replacement, there is an opportunity to move away from traditional preventative maintenance and shift to predictive maintenance. With predictive maintenance, significant reductions in unplanned downtime can save millions, keep planes flying and keep customers happy.

3 Data

What if invisible insights into your business became visible? What if terabytes of data pulsing through your operations were captured and stored securely in the field and in the Cloud to be accessed in real time? Imagine an enterprise world with a single source of truth—a panoramic view of your entire fleet and operation with the ability to zoom into the details in a cost-effective way, and a workforce focused on resolving issues, innovating and collaborating across silos. Data—and more importantly, data science—are the keys to unlocking this potential, and can make predictive maintenance not just a reality but a pervasive reality.

In aviation, data volumes are growing exponentially—an Airbus A350 generates and archives 50 times more data than an Airbus A320; from 12,000 parameters and 8.3 TB to over 670,000 parameters and 450 TB. Airline, OEM and MRO organizations currently struggle to store, manage, use and understand this flood of information. When the data is stored and managed, it is typically stored in data warehouses that are difficult to access or query by business users, making it impossible to apply data science and analytics to create value. That's why an open Big Data platform is an essential component to a maintenance planning system.

To handle these massive data sets, you need a platform for connecting, securing and analyzing data. In 2012, GE began developing their solution, an open, Cloud-based software platform named Predix that could provide machine operators and maintenance engineers with real-time information to schedule maintenance checks, improve machine efficiency, and reduce downtime.

It didn't take long for developers of General Electric's Industrial Internet platform, Predix, to realize that they could find interesting and unique patterns in the data. They thought the patterns of sensor data could be used to provide an early signal of future performance problems and better predict when its machines should be scheduled for maintenance. In early 2013, GE began to use Predix to analyze data across its fleet of machines.

GE noticed that some of its jet aircraft engines were beginning to require more frequent unscheduled maintenance. GE used data science techniques to build transfer functions that use full flight data from a small set of flights to model and predict for the larger fleet. That data coupled with environmental, air quality, city pair and other data allowed GE to expand its' ability to build complex, multi-variable predictive models to accurately segment beyond what traditional physics-based modeling can do.

By pulling in massive amounts of data and using fleet analytics, GE was able to cluster engine data by operating environment. The company learned that the hot and harsh environments in places like the Middle East and China clogged engines, causing the high pressure turbine shroud to heat up and lose efficiency, thus driving the need for more maintenance on planes that passed through. All of that was done because GE could use data from GE engines across the world and cluster fleet data. The results, the first ever analytics based AMOC (Alternative Method of Compliance) was delivered to a targeted population of engine operators, thus saving most operators from having to perform unnecessary preventative maintenance. Huge win for airlines, MROs, GE and all of their customers; all of it was made possible through an open, robust data and analytics platform.

4 Analytics

The key to delivering insights from data and driving predictive maintenance actions lies in analytics. By analyzing what differentiates one machine's performance to another, what makes one more efficient, for example, you can more tightly hone its operational parameters (Winig 2016).

In the previous high pressure turbine shroud example, an analytic approach was developed that allowed us to understand the exact transfer function between the part-life, engine environment and duty cycle of operation. GE was able to segregate the fleet and characterize the status of each engine. Using the analytic approach, improvements including climb de-rate and an optimized water wash process were implemented at customers to reduce the distress on a specific part. These same operational improvements also reduced fuel burn and saved one customer $7 million annually in fuel cost. So, in addition to prioritizing specific engines in terms of their risk for an early removal or service disruption, GE was also able to provide guidance on operating parameters, or behavioral changes, to mitigate or abate the risk of early removal. Again, a predictive maintenance win. Fig. 1.

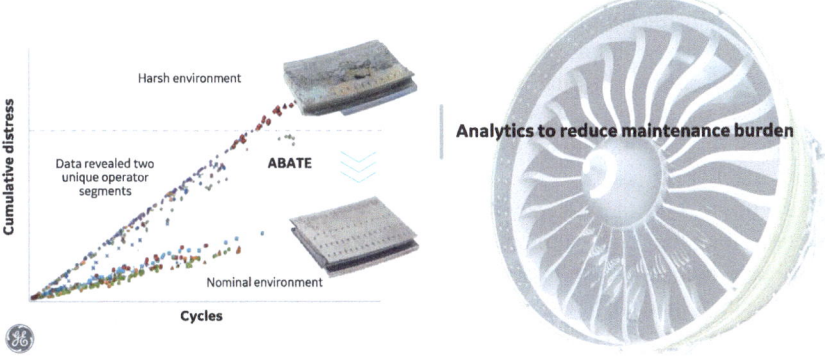

Fig. 1 Analytics to drive predictive intervention

More data and better analytics allow airlines to have more accurate intelligence on when an aircraft or its parts will need service or repair and what types of maintenance issues are likely to arise, thus affording the luxury of time. Time to schedule the work at a time and place potentially more convenient and more economical than if the repair was unplanned. And since predictive maintenance is more precise than average-based or historically based preventive maintenance schedules, it is more effective. Serviceable parts won't be replaced unnecessarily simply because they meet historical criteria.

To realize maximum performance and efficiencies, organizations serving airlines must do more than capture and manage data. MROs must deploy data analysis tools to enable real-time analytics and build predictive models. This gives you real insights into key processes and variables, while anticipating and planning for events.

New data analytics tools give regular business users better insights but, to get the greatest value and most useful insights from available data, you need highly qualified staff of data scientists who know how to use Big Data. For most organizations, this means recruiting new, skilled employees or engaging a partner that has the talent you need.

5 Digital Twin

The ultimate application of data and analytics is with the digital twin. The digital twin represents all of the digital information collected about a specific physical asset. We collect its history, the conditions under which it has been used, its configuration, its maintenance history and more. The idea is to maximize the life of the individual asset, optimize its performance, and optimize a set of assets across a fleet - all through advanced analytics.

The relationship between the worker and the machine is symbiotic. Most fundamentally, the machines need to be designed to send the signals that will let humans use them more effectively and the machines need to "read" humans more effectively. In the meantime, design needs to be simpler and focused on adapting to what people need, so employees can focus on business outcomes and impact. The digital twin is meant to enable workers to be more effective, productive and important to the business. It is about augmenting and enhancing the worker's capabilities, and enabling them to keep up with the rate of change and leveraging it more effectively.

In the future, human workers will be collaborating with intelligent devices more than controlling them, and will be supported by intelligent agents. They'll be using model-based human-system interfaces that let them focus on business transforming decisions rather than having to deal with the increasing complexity and scale of the systems they are using. The digital twin is the black box behind the applications that will help airlines, OEMs and MROs connect machines and humans.

The digital twin is a collection of physics-based methods and technologies that are used to model the present state of every asset in an airline. The models start by providing guidance on "design limits." The models are continuously updated, and learn to accurately represent the asset under a large number of variations related to operations—fuel, temperature, air quality, moisture, load, weather forecast models, and more. Using digital twin models and state-of-the-art techniques of optimization, control, and forecasting, the applications can more accurately predict outcomes along different axes of performance, reliability and maintenance. The models in conjunction with the sensor data give us the ability to predict the assets performance, evaluate different scenarios, understand tradeoffs, enhance efficiency, and ultimately drive predictive maintenance.

6 Intelligent Machines

The rapid decline in the costs of both electronic sensors along with data storing and processing—thanks to advances in Cloud computing—now allows us to harvest massive amounts of data from industrial machinery.

Thanks to an increase in sensor-enabled aircraft—fueled by a decrease in the price of sensors—predictive maintenance is growing in acceptance and use in the aviation industry. A sterling example of the growing influence of Big Data and the Industrial Internet: predictive maintenance leverages these evolving technologies to take proactive maintenance, and the aircraft reliability and safety it offers, to an entirely new level.

Machines like gas turbines, jet engines, locomotives and medical devices are becoming predictive, reactive and social, making them better able to communicate seamlessly with each other and with us. The information they generate becomes intelligent, reaching us automatically and instantaneously when we need it and allowing us to fix things before they break. This eliminates downtime, improves the productivity of individual machines—as jet engines consume less fuel and wind turbines produce cheaper power—and raises the efficiency of entire systems, reducing delays in hospitals and in air traffic.

Machines will play an active part in this; connected and communicative machines will be able to self-monitor, self-heal, and proactively send information to other machines and to their human partners. Intelligent machines are securely managed industrial devices that autonomously connect to the Industrial Internet, execute native or Cloud-based machine apps, and analyze collected data and react to changes in the data. They are predictive (anticipating and reacting to state changes), reactive (sensing the environment and acting on it), and social (communicating with each other and other industrial resources).

While many industrial assets have been endowed with sensors and software for some time, software has traditionally been physically embedded in hardware in a way that the hardware needs to change every time the software is upgraded. We are beginning to deploy technologies like embedded virtualization, multi-core processor technology and advanced Cloud-based communications throughout the industrial world. This new software-defined machine infrastructure will allow machine functionality to be virtualized in software, decoupling machine software from hardware and allowing us to automatically and remotely monitor, manage and upgrade industrial assets. In other words, the software of new-generation industrial assets can be upgraded remotely without changing any hardware at all—much like the software in our smartphones, instantly enabling new functionalities.

Software that understands a machine's physical capabilities relative to its theoretical potential can do more than simply detect variance; it can adjust operating parameters in real time, maximizing efficiency and minimizing cost, and prescribe predictive maintenance actions to reduce and/or eliminate unplanned downtime.

7 Supply Chain

In manufacturing, predictive maintenance is designed to transform real-time machine data into actionable metrics to boost productivity by preventing or minimizing unscheduled downtime; making time-based preventive maintenance a thing of the past.

Operational platforms that identify maintenance issues can proactively transfer data to IT systems that manage maintenance personnel and supply chain vendors, ensuring that the necessary parts are replaced by the right personnel at the right time. Through the use of analytics and a holistic perspective, airlines, OEMs and MROs can optimize their performance and improve their fleets' reliability while minimizing maintenance costs.

In early pilot tests, agricultural machinery manufacturer CNH Industrial N.V. found that switching its internal paradigm from preventative to predictive reduced average downtime for combines, tractors, and harvesting machines by approximately 50%. The company plans to leverage Big Data for additional value, for instance, developing the machine intelligence to know that when the water temperature on a particular model rose to a certain level, 80 % of the time the vehicle was experiencing a faulty radiator. With this knowledge, the parts supply chain could ensure the nearest dealer has the necessary parts for replacement.

The reality for the MROs of a mixed fleet can be daunting but Big Data and analytics can help. The global commercial aviation sector is growing, but it continues to struggle with costs, profitability, changing customer expectations, and regulatory burdens. Experts project the world airliner fleet to grow from approximately 25,000 today to 44,500 by 2033. New aircraft types will have longer planned maintenance intervals, while the existing fleet will stay in service for up to 20–25 years, with many aircraft seeing longer use if they are converted to freighters. Even with longer maintenance intervals on newer aircraft, the pressure across the supply chain for greater dispatch reliability, increased annual availability, and lower maintenance costs will continue. These efforts can greatly benefit from data analytics.

One vision for how the Industrial Internet can impact the aviation supply chain comes from the area of aircraft maintenance inventory management. An intelligent aircraft will tell maintenance crews which parts are likely to need replacement and when. This will enable commercial airline operators to shift from current preventative maintenance—based on the number of cycles—to predictive maintenance schedules that are based on actual need. The combination of sensor data analytics, and data sharing between people and machines is expected to reduce airline costs and improve maintenance efficiency.

Big Data analytics can help identify risk of future failure based on sensor feeds but, additional context, such as engineering data and past maintenance history—for the part in general and for the specific aircraft—make the analysis much more meaningful. Once the predictions are in hand, aircraft builders, operators, maintenance professionals and logistics providers can combine to make up a truly effective predictive maintenance program that reaps maximum value.

8 Better Together

As we have outlined above, digital technologies are set to completely transform the entire aviation industry value chain. This will require airlines, OEMs, MROs and suppliers to adapt and adopt new business models, while emerging players introduce new services and capabilities.

This may require changing their business model to fully take advantage of new digital capabilities. A first priority will be to use the insights provided by Big Data analytics in order to ensure and increase asset reliability. Conventional preventative maintenance procedures will remain a vital component of fleet maintenance, but new technologies will accelerate the adoption of predictive maintenance, requiring software to manage and optimize the fleet reliability. Switching from preventive to predictive maintenance to maximize uptime will also require a transformation. In general, airlines, OEMs, MROs and suppliers across the value chain will have to adopt a "data first" mentality, always thinking in terms of the potential insights that can be gleaned from data and analytics to improve the value of the service.

This transformation will need a high degree of coordination among stakeholders. Airlines, OEMs, MROs and suppliers will need to work together to realize the outcomes of reduced costs, increased efficiencies and improved reliability. These participants will act like virtual predictive maintenance teams, determining the status of the aircraft and its subsystems to supply real time, actionable information to help operators predict failures before they occur and provide quick and accurate "whole plan" views of asset health and reliability. The opportunity for all participants in the future of aviation is unprecedented—it is full of digital promise.

9 Outcomes Delivered

9.1 Asset Performance Management

One of the clearest and greatest benefits of Industrial Internet solutions is substantially reduced equipment downtime through predictive maintenance. Advances in predictive analytics can now indicate when a piece of equipment is likely to experience a specific failure, so that maintenance can be performed ahead of time. This prevents the failure and resulting downtime, and allows time to efficiently adapt operations in cases where the asset needs to be taken off-line for maintenance. Managing assets in the future will be software defined.

9.2 Maintenance Optimization

Optimal, low-cost, machine maintenance across fleets can also be facilitated by intelligent systems. An aggregate view across machines, components and individual parts provides a line of sight on the status of these devices and enables the optimal number of parts to be delivered at the right time to the correct location. This minimizes parts inventory requirements and maintenance costs, and provides higher levels of machine reliability. Intelligent system maintenance optimization can be combined with network learning and predictive analytics to allow engineers to implement preventive maintenance programs that have the potential to lift machine reliability rates to unprecedented levels.

9.3 Operations Optimization

The largest value-creation opportunity lies in operations optimization, specifically creating an integrated approach to network optimization for airlines and airports. Today's typical operations landscape consists of a high degree of fragmented point solutions that each satisfies a specific operational need. There may be a system that manages engine and aircraft data capture and transmission, for example, another system that manages date of operations and another for monitoring and diagnostics of assets, and so on. However, the real value-creation opportunity exists in connecting these silos and driving collaboration across airlines, airports, OEMs, MROs and suppliers.

9.4 The Power of 1 %

From an operations perspective, the average cost of maintenance per flight hour for a two engine wide-body commercial jet is approximately $1200. In 2011, commercial jet airplanes were in the air for 50 million hours. This translates into a $60 billion annual maintenance bill. Engine maintenance alone accounts for 43 %of the total, or $25 billion. This means that commercial jet engine maintenance costs can be reduced by $250 million for every 1 % improvement in engine maintenance efficiency due to predictive maintenance.

9.5 Analytics-Driven MRO

Strategic benefits of data driven MRO are readily apparent. Better information and predictive information let the MRO improve their overall efficiency through optimized staffing, planning, sourcing, inventory management and repair activities. Predictive analytics can significantly reduce unplanned maintenance by enhancing failure prediction, accelerating root cause identification, accelerating incident response and optimizing inventory management and allocation.

9.6 Reduced Downtime

Ten percent of flight delays and cancellations are currently caused by unscheduled maintenance events, costing the global airline industry an estimated $8 billion—not to mention the impact on all of us in terms of inconvenience, stress, and missed meetings as we sit helplessly in an airport terminal. Predictive maintenance is used to address this problem. While in flight, the aircraft will talk to the technicians on the ground; by the time it lands, they will already know if anything needs to be serviced. For U.S. airlines alone, this system could prevent over 60,000 delays and cancellations a year, helping more than seven million passengers get to their destinations on time.

10 Conclusion

Blending the physical and digital has led to a world where humans and machines communicate and collaborate. The proliferation of Cloud and automation technologies will lead to more seamless and real-time simulation, workflow, and collaboration to transform the design of new products and accelerate the development process.

Predictive maintenance is a defining benefit of the Industrial Internet. Digital tools will track and maintain historical performance baselines for individual assets as well as the entire fleet, comparing it to real-time performance monitored on a continuous basis. Any variance from "expected behavior" derived from these baselines or expected operation will trigger an alert and possible action. Advanced analytics will then determine whether the variance signals a potential future malfunction, its root cause and the likely timeframe over which the malfunction will occur. Cost-benefit analysis of how much longer the asset can perform, and at what load, before the issue must be addressed will become the norm. This will allow the airlines or MROs to address issues proactively, reorganize workflow around planned maintenance and avoid unplanned downtime.

This will bring us closer to a world of no unplanned downtime, no maintenance-related delays and cancellations, and no aircraft stranded on the ground for mechanical failures. It will dramatically improve capacity utilization, and reduce the time we lose today by performing preventive maintenance and servicing for lack of information on the actual status of the assets.

The key lies in realizing the digital future of the aviation industry. Industrial Internet solutions enable a shift to optimized performance and predictive maintenance, yielding major cost savings and profitability for all who participate. Benchmarking across fleets and operators will now become possible, as data is transmitted back from assets to be aggregated and analyzed. Operational anomalies can be identified and corrected, and airlines that perform better than expected can be rewarded.

References

Winig L (2016) GE's big bet on data and analytics. MIT Sloan Management Review. Available via http://sloanreview.mit.edu/case-study/ge-big-bet-on-data-and-analytics/. Accessed 22 Jun 2016

Jim Daily, is Vice President and Chief Digital Officer for GE Aviation. Jim was promoted from Vice President Engineering and Technology to Chief Digital Officer when GE Aviation formed a digital organization in early 2016 that brings all of the digital expertise from across GE Aviation into one business. In this role, Jim is responsible for the technology and business growth for the digital platforms and portfolio within GE Aviation.

Jim joined GE in 2011 after a 22-year career at Honeywell in several engineering and business roles of increasing responsibility culminating is his role as Vice President, Electronics Center of Excellence. In conjunction with this role, he led the Aerospace Integrity and Compliance Council and was honored as the 2009 recipient of the highest corporate award for product development. Jim is a graduate of California State University at Fullerton, where he earned a Bachelor degree in electrical engineering, with an emphasis in computer and micro-processor based systems and digital design.

Jeff Peterson, is the Digital Marketing Director for GE Aviation where he is responsible for the digital strategy, value proposition and go-to-market for digital platforms and portfolio within the aviation industry. Prior to that, Jeff spent 18 years with GE where he held positions as Chief Marketing Officer of Taleris (GE Aviation and Accenture joint venture), Marketing Director of GE Advanced Sensors, Marketing Program Manager, Sales Operations Manager, Black Belt, Project Manager and Manufacturing Engineer with GE Energy/Oil and Gas. Jeff earned a Bachelor degree in Mechanical Engineering from Montana State University and a Master in Business Administration from the University of Nevada Reno.

Outlook

Stefan Berndes

Abstract The German Aerospace Industries Association BDLI is the voice of the country's aerospace sector. Working closely with its members and drawing on their expertise, BDLI studies the long-term development of the industry. The chapter outlines the association's assessment of the current debates on the most salient issues with a particular focus on the supply chain. Hence the ongoing strategic re-orientation of the German supply chain with regard to OEM industrialization and internationalization strategies presents one important aspect. This materializes with regard to the cabin interiors sector which is quite important for the German aerospace industry in a Vision Cabin/Cargo. Furthermore the text dwells into the dispute of digitalization and using a German term "Industrie 4.0" showing the specificity of the implementation of appropriate IT-solutions in this high-tech, safety dominated, low volume sector. Such issues are expected to shape the future of the industry. Based on a number of examples, market forecasts and studies, this chapter aims to highlight the most important developments and trends that may impact companies in the sector and advocates an active approach to shape its future which is laid down in the current aerospace technology roadmap.

1 Introduction

Providing an outlook on the future of the German aerospace industry has become a mainstay of our association. We naturally focus on our home market although the German aerospace industry is an integral part of Europe's aerospace industry, which in turn is becoming more and more global.

Our analyses focus on relevant long-term business and technological developments that are evaluated in light of the wider international socio-economic context. Assessing industrial structures and processes in the aerospace supply chain is an important part of this work. While we naturally tend to highlight positive

---

S. Berndes (✉)
Bundesverband der Deutschen Luft- und Raumfahrtindustrie e.V., Berlin, Germany
e-mail: berndes@bdli.de

developments to our stakeholders, we also consider it our responsibility to equally consider the risks of future developments.

This chapter aims to highlight key areas currently debated in the sector and within our association. We expect these developments to shape the future of our industry, not least the aerospace supply chain. Given its importance to Germany's aerospace industry, naturally the emphasis of this chapter is on civil aviation.

The German Aerospace Industries Association (Bundesverband der Deutschen Luft- und Raumfahrtindustrie e.V.—BDLI) represents the interests of more than 230 members. Based on its technological leadership, Germany's aerospace industry is an international success story which has become a significant driver of economic growth in the country. With over 106,800 employees and a turnover of 34.7 billion euros in 2015, German aerospace companies produce all key technologies for this strategically important sector.

Communication with political institutions, authorities, associations and foreign representations in Germany is a major task of the BDLI, as well as a variety of services in Germany and abroad for its members. BDLI also owns the trademark rights to the ILA Berlin Air Show International Aerospace Exhibition.

2 The Future May Hold: The Standard Scenario

The two leading aircraft manufacturers Airbus and Boeing regularly try to answer the question: "What does the future hold for the aeronautics industry?" Their prognoses tend to be the source for our own deliberations. We refer to their take on the future as the "standard scenario".

Boeing's "Current Market Outlook" (Boeing Commercial Airplanes 2015) forecasts a demand of more than 38,000 aircraft in the 2014–2034 period to meet a projected global increase in passenger and freight revenue kilometers. Similarly, in its annual "Global Market Forecast" (Airbus 2015), Airbus expects a demand for almost 33,000 aircraft during the same period.

The variance between these two annually published studies is mainly due to a divergent view on the airline needs over the coming decades. While Boeing anticipates more direct links between city pairs, Airbus expects the hub-and-spoke model to become more economically stable and profitable for airlines.

What they have in common, however, is the expectation that the well-understood link between economic growth and air traffic development will, by and large, continue to persist well into the future (Figs. 1 and 2). This may well turn out to be accurate, given that large developing economies such as China experience significant economic growth, and their emergent middle classes can be expected to generate a growing demand for business and leisure air travel for some time to come. Furthermore, the manufacturers' forecasts assume that external shocks such as SARS or terrorist attacks will have no long-term effect on growth, and that the price of kerosene will not rise rapidly.

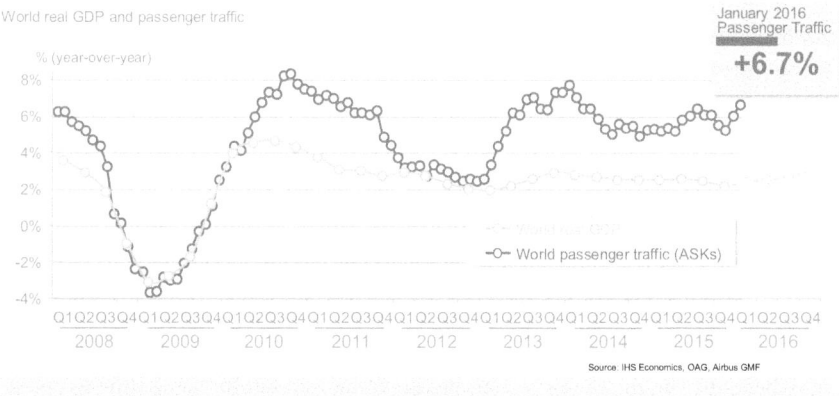

Fig. 1 Market development PAX versus GDP

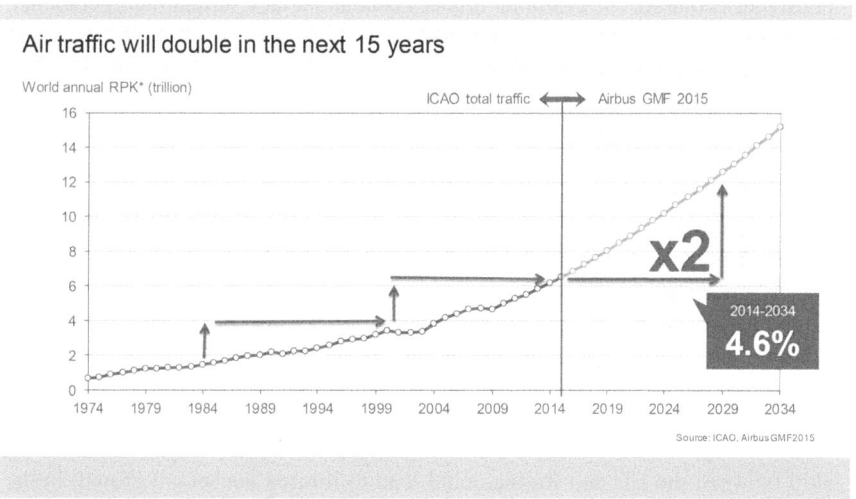

Fig. 2 Market development forecast air travel

3 How the World Might Develop: Alternative Scenarios

It is a well-known truism that the future is unknown and might turn out to be different from what is generally expected. To understand how the future of the aeronautics industry may differ from current business-as-usual assumptions, our association studied various scenarios. The analysis revealed three main issues that could shape the global environment for the aerospace industry until 2035:

- A growing gap between energy demand and energy supply (the energy security question);
- Environmental issues, in particular climate change, caused by CO_2 emissions. As one expert put it: "There will be a price on CO_2 in 10 years' time"; and
- Geopolitical changes due to demographic and economic factors, leading to a shift of hard and soft power to Asia.

In view of these key issues and various other factors, we developed three scenarios for the German aerospace industry:

1. Under the first scenario, which we call "a world of global players", market liberalization remains the dominant force in international trade, and measures to address climate change are introduced too late. Hence, toward the end of the 20-year period in question we could see drastic regulatory measures to curb growth across the board. Under this scenario, the German aerospace industry may benefit from the sector's continued revenue growth for some time to come, provided it manages to participate fully in the global supply chain of the leading manufacturers. The main threat is a potential loss of technological leadership in strategically important areas, as well as a deterioration of the industry's competitiveness due to, for example, higher costs and a lack of qualified personnel. Capabilities in military aviation could be lost completely because of further European integration. Under this scenario, the biggest threat to the sector is posed by a future environmental crisis which might put a stop to the development of the civil aviation industry, unless "green technologies" will be introduced in time.
2. An alternative is offered by the second scenario, entitled "struggle for resources in a multipolar world". It assumes the emergence of geopolitical conflicts triggered by Asia's growth and resource hunger. China's ascent and its run on Africa's resources could be a precursor for such conflicts. The result may be protectionism, a reversal of efforts to liberalize trade, and a low degree of international cooperation. Environmental issues also play a decisive role, but in different ways than in the first scenario. Rather than international action to curb emissions, this scenario predicts limited measures taken at the national level to adapt to a changing environment. Furthermore, an antagonistic relationship between the EU and Russia could lead to limited but secure growth in the strategically important aerospace sector, even if new competitors emerge from Russia. Manned space missions could benefit from the prestige that nations gain from this activity. "Corrective" environmental technologies which would enable societies to adopt to a changing environment could become an opportunity for the German aerospace industry. However, towards the end of the period, resource and environmental issues could lead to a stagnation of air transport.
3. A different take on the rise of Asia is offered by the third scenario named "megacities drive technology". The increase in the number and size of highly dynamic megacities in emerging markets could lead to new innovative solutions addressing the most pressing transport challenges of our time. These might differ substantially from the solutions sought out in Europe or the US due to path

dependency in the West. Megacities in India and China could, for instance, decide to prioritize ground transportation systems such as high-speed rail over aviation. This could lead to a lower than expected demand for short-haul aircraft. Nevertheless, significant opportunities exist under this scenario, for instance, for long-haul aircraft to provide mass transportation between a growing number of megacities, or for helicopters to avoid the congested roads of metropolitan areas. It is worth noting that this scenario poses a significant threat to the German aerospace industry with its focus on short and medium-haul aircraft. On the other hand, the megacities of the future with all their potential conflicts and challenges might benefit from civil and military aviation technology to provide a high level of security to their citizens. In addition, aerospace and space exploration in particular might be seen as a sign of hope in a crowded world. However, crowded cities might also find different ways to provide services such as Internet access provided by drones built by technology companies that traditionally fall in the remit of the aerospace industry such as telecommunication and observation via satellites.

In summary, the future may turn out to be different from the standard business-as-usual scenario. Considering the aerospace industry's need to further develop robust supply chains and industrial systems, in the following section we will discuss how the industry should react and how an international division of labor might look.

4 Shaping the Future

There are good reasons for Airbus and Boeing to believe in their standard scenarios. Based on the aerospace industry's experience gathered over the past 50 years, it is assumed that a number of well-understood trends will continue for at least two more decades without major upheavals. And given the long development times and lifecycles of any aircraft model, 20 years is not even a particularly long timeframe. In addition, the expectation of a stable future provides stakeholders—be they shareholders, suppliers or political decision-makers—with a degree of dependability they require to take informed investment decisions.

Having said that, if the aerospace industry wants to see the standard scenario become reality, it will have to actively shape the future. This insight highlights some important a.o. moral obligations, in particular that the industry should only attempt to shape the future in ways that produce outcomes in line with today's ethical norms. Simply put, the challenge is to ensure the industry remains successful while avoiding outcomes that negatively impact world populations.

Which areas does this concern? A key task is to continue to advance the industry to meet demand. Airbus currently sells more than 1000 aircraft and builds more than 630 aircraft per annum (Airbus 2016). In view of the considerable downward cost pressure, the structure of the industry and the technological requirements, scaling up production is a tall order. In addition, countries which place large orders increasingly expect to participate in the industry's value and supply chain.

4.1 Studying the Supply Chain

The German Aerospace Industries Association has carried out a number of surveys to assess the supplier industry's structural adjustment to constantly evolving challenges. These studies have consistently highlighted the need for the German equipment and materials industry to rise to the challenges brought about by the global expanding aeronautics industry market. These include global competition, raising capital to finance growth, and establishing an appropriate market presence and efficient sales structures in growing markets such as China and Russia. For these reasons, the pressure to consolidate remains high.

Regarding the German equipment industry, globally operating, well-funded corporations including Zodiac, Thales, Cobham, and Smiths have emerged in neighboring countries. These companies expand their product portfolios and technological abilities by means of acquisitions and by heavily investing in Research and Development. Unsurprisingly, they have managed to increase their market shares. As far as Germany is concerned, strategic consolidation has been limited to defense technology (Airbus), ground systems and flight control (Liebherr Aerospace), and certain areas of civil avionics (Diehl). At the same time, foreign companies have been 'cherry picking'. Several German equipment and materials companies have been absorbed into large international corporations (such as SEL by Thales, and Nord-Micro by UTAS). One group of companies or specific operations has been identified where a consolidation process was a promising option. This was the industry specialized in cabin interiors. The following chapter will give an overview of the current state of play.

The remaining equipment industry in Germany is fragmented and largely consists of medium-sized companies. It is in danger of slipping even further down the supply chain, as its lack of size and competitiveness continues to reduce its role in large international programs. The result from 2005 is *cum grano salis* valid up to date.

The remaining independent German aerospace companies are unlikely to become consolidated in any competitive national alliance since they are too small, do not produce the same or complementary products, or manufacture tier 2 and tier 3 products. Although the German companies that have already been consolidated have good prospects of holding their own on the world market, they too are faced by the serious challenge of keeping up with the rapid pace with which globally aligned companies are increasing their market shares. Companies bought up by foreign groups have developed well, mostly thanks to their integration into powerful financial structures.

4.2 Catalyzing Cabin Industry

Airlines increasingly use the aircraft cabin to differentiate themselves from their competitors. Cabin interiors are updated and upgraded more regularly, and advanced technologies have opened up new possibilities to markedly improve the passenger experience. Hence the market for passenger and also freight cabin systems is growing even faster than the aerospace market on the whole.

However, the resulting growth potential can only be realized if manufacturers and airlines were willing to source cabin and freight modules from suppliers producing high-quality, low-cost, integrated modules or systems as part of a tiered supply chain. This constellation is both an opportunity and a challenge: companies that want to be seen as viable module or system suppliers have to take on some of the integration, engineering and development tasks of aircraft manufacturers. Moreover, they have to manage the downstream supply chain and the resulting complexity, while at the same time meeting ever-higher expectations. Despite these challenges, in the framework of BDLI a strong cabin cluster has emerged in Germany in recent years mostly due to the now well-positioned Diehl Group, which acquired a cabin production site from Airbus, lavatory manufacturer DASELL, and galley producer Mühlenberg (now Diehl Comfort Modules and Diehl Service Modules). At the same time, ZODIAC Group took over Sell, the leading German galley manufacturer.

Consequently, in the area of cabin manufacturing, a strong supplier industry has developed in Germany which is now also being recognized in the political arena. The Federal Government's Aviation Research Program, for instance, has acknowledged the importance of the cabin industry, and Hamburg—home to the world's third largest aircraft production site—has made this area one of the focal points of the recently launched aviation research center ZAL.

4.3 Foresight: Vision Cabin/Cargo

Building on the above-mentioned surveys of the supplier industry, our association and its members engaged in more in-depth scenario planning. Most importantly, under the project name "tomorrow's flying experience", the first joint industry vision for the aircraft cabin of the future was developed. The main ideas and results were summarized in the so-called "Vision Cabin/Cargo" (Fig. 3).

This vision is an attempt to answer the question: "What does tomorrow's flying experience look like, and how will it impact the look and feel of the aircraft cabin of the future?" The vision highlights six evolutions (Max Space Architecture, Seating Solutions, Storage Management, Connectivity Cloud, Active Surfaces, Hygiene Lavatory) of in-flight innovations which have been identified as essential for future cabins. However, the cabin vision should not be understood as the future cabin per se, but as a platform where ideas and future market expectations collide.

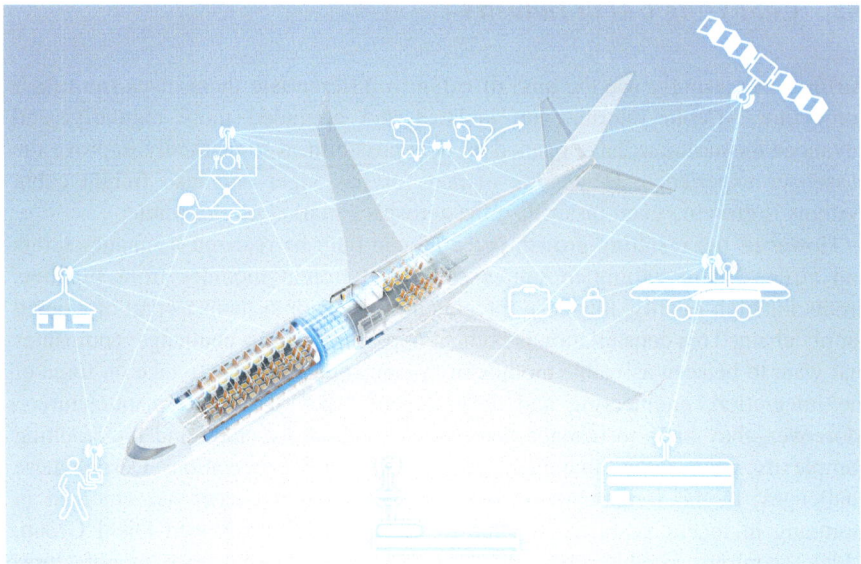

Fig. 3 Vision cabin-cargo

Our network identified a number of core technologies which are now being further developed by various industry players. The following provides an overview of six evolutions as well as current and future R&T topics that may well shape the passenger experience in future decades.

Max Space Architecture
A multi-level architecture in the cabin increases efficiency and reinvents flexibility and differentiation. X-flexibility: cabin space can be adapted to bookings, in other words: high versus low density pitch could be associated with pricing. Double deck architecture: a folded floor ensures full standing height within a minimum cross-section. The systems are located on crossing between floor and lateral fuselage. Y-flexibility: seating benches adapting to individual needs, ergonomics and business models create a new individualized products and services scope. A wide spectrum of configurations reach from high density seating to high premium seating.

Seating Solutions
Depending on the layout, there is potential for a 20–30 % increase in passenger capacity in the double-level zone (compared to the single deck layout).

Up till now, since the aircraft layout is not specified, the total number of seated passengers can only be estimated. However the concept reveals a significant increase in seat-count compared to existing single aisle programs.

Storage Management
The positioning of hand luggage in the central boarding zone enables a smart and efficient usage of cabin space and improved cabin processes for boarding/deboarding.

Luggage revolver: a circular arrangement improves the use of idle space; an automated rotating system offers more dedicated storage space to the passenger than available today. Access to hand luggage during the flight is guaranteed.

Loading station: automated luggage access based on radio frequency identification (RFID) saves time and reduces passenger effort. Simplified handling with ergonomic loading position provides direct access from inside at any time. Loading and unloading is also feasible from outside the aircraft.

Connectivity Cloud
The permanent connection to a global data network enables individualized traveling experience and continuous access to online services, as well as efficient aircraft and cabin operations.

Smart products and services: seamless passenger locating and guiding at the airport and in the cabin ensures carefree travelling.

Continuous luggage monitoring improves efficient aircraft operations and minimizes loss.

Entertainment and communication are improved by real-time personalized IFE systems. Contents can be individually chosen from the Cloud and will be displayed on smart interactive surfaces.

Catering content management will be customized according to operator and passenger choices.

Active Surfaces
Cabin modules with integrated system functions and standardized interfaces can be positioned flexibly throughout the aircraft surface to improve operational and resource efficiency.

Multifunctional surfaces: an individualized cabin environment and cabin appearance will enhance passenger experience. Displays, cabin lighting, signs and comfort functions will be part of a programmable display matrix and adjusted according to any seat configuration.

Sensor technology and oxygen supply: a network of sensors and a smart adaptable emergency system provide cabin safety.

Insulation and systems: adaptive and highly efficient insulation materials with integrated systems will increase acoustic and thermal passenger comfort.

Hygiene Lavatory
The hygiene lavatory focuses on sensual perception (see-hear-touch-smell) to give an experience of cleanliness and individual service.

Smooth surfaces and a self-cleaning floor improve cleaning procedures. The seat disinfection guarantees a sterile impression.

Innovative lighting: mood lighting intensifies a perception of cleanliness. UV-lighting increases the visibility of amenities and water.

Acoustics: quiet ambiance and pleasant sounds for a relaxing and comforting atmosphere.

Fragrances refresh the space.

Interaction: sensor-activated automatic and touchless interfaces ensure carefree handling.

Spaciousness and customized comfort: passengers can choose individual scenic wall projection displays to improve the overall lavatory atmosphere.

5 Technology: The Mean to Shape the Future

Technology is a key component that will turn our visions for the future into reality. Superior technology is a competitive advantage in all parts of the aeronautics value chain, as is a low-cost base due to the low margins in our sector. Considering these two somewhat competing factors, technology is the key to remain competitive, in particular for a high-wage country such as Germany.

5.1 Improving the Supply Network: Industry 4.0

Reliance on superior technology is particularly important when considering "Industry 4.0". While the concept itself is still vaguely defined, the aeronautics industry is actively discussing the subject. Germany's Federal Government has already prioritized Industry 4.0. A few definitions help to understand the Industry 4.0 concept:

1. pwc's Strategy& (Koch et al. 2015): Fourth industrial revolution (pwc/strategy&) is defined by increasing digitalization and networking of products, value chains and business models with data access and data analysis as core competences.
2. Roland Berger (Berger Strategy Consultants 2015): Digitalization of Industry through 4 levels of transformation: digital data, automation, networking and digital customer access.
3. T-Systems (T-Systems 2016): Industry 4.0 is a project of the high tech-strategy of the German Federal Government. It facilitates the digitalization of established industrial sectors. The aim is to arrive smart factories which are distinguished by flexibility, resource efficiency and ergonomics and not least integration of customers and suppliers into business processes and value chains (T-Systems).

Industry 4.0: Classification and Evaluation
We remain skeptical whether the defined changes of the Industry 4.0 approach will result in an "industrial revolution". If, for example, Industry 4.0 is to be classified as the 4th industrial revolution to have taken place over the last two centuries, a middle-aged person would have experienced it least the last two revolutions. And how can such an open concept represent a revolution?

Nevertheless, Industry 4.0 represents opportunities to expand and improve the aerospace business. At the same time, it poses a prominent threat to the established aerospace business ecosystem.

There are at least three aspects to be considered carefully by our sector:

1. One string of debate is tied to the concept of Computer Integrated Manufacturing (CIM).

   CIM is the manufacturing approach using integrated systems and data communication, coupled with new management philosophies that improve organizational and personnel efficiency (Wikipedia 2016). Technology and tools are currently available to employ CIM, as well as even more cutting-edge production concepts. Over the past two decades, the aerospace industry's systems have been equipped with data processors, sensors and actuators throughout nearly the entire production process, and we also benefit from the ubiquitous availability of broadband and high-performance data communication. In addition, new technologies, such as data goggles, cooperative robots, and additive layer manufacturing (ALM) technology, are beginning to leave their mark.

   Still, the challenges are enormous considering data integration, design, simulation, product and production data modeling (ERP, CAD, CAM etc.). CIM takes into account: relevance, traceability, acceptance, consistency, coherence, cost-efficiency, security and fault tolerance.

   However, significant savings from a reduction of lead times, quality improvements and the optimization of the use of working capital are to be expected. And more than the CIM-concept of the 1990s, which concentrated on the integration of data model of a single organization, the current Industry 4.0 concept has expanded to include multi-tier supply chains.

   Furthermore, Industry 4.0 may trigger a change in the supplier structure towards more flexible supply networks instead of chains. That may allow different and better cooperation across the suppliers and vis-a-vis Tier 1 and OEMs. As such it might become a great chance particularly for the German suppliers to overcome the current small-size and non-cooperation situation while becoming as efficient as OEMs request. However, this Industry 4.0 vision still has to be further elaborated and adapted to the current structures.

2. A second strand of the debate encompasses new business models and opportunities for businesses. The once world-leading European mobile phone industry with players such as Nokia and Siemens, for instance, was rendered meaningless over the past decade by companies such as Apple and Samsung. According to management consultancy Roland Berger, the reason was a transformation of the customer interface which allowed Apple to grow its proprietary ecosystem of hardware and software. Even if this process may be not as applicable to the aviation business because customer interfaces will not really change significantly for the OEM business and market entry remains high, it might change parts of it. The MRO business, for example, has to adapt if additive manufacturing allows to produce spare parts around the world at any time and OEMs like

engine or landing gear manufacturers sell their equipment based on flight hour agreements or cycles.
3. Any established sector tends to look with concern at the vast amount of capital accumulated by technology companies like Google, Amazon and Facebook. This capital can—and is—being used to enter a wide variety of markets. To quote Samuel Beckett's Worstward Ho: these companies "...Ever tried. Ever failed. No matter. Try again. Fail again. Fail better." As an example, Elon Musk and Amazon founder Jeff Bezos are trying to redefine the space business with their creations SpaceX and Blue Origin.

What does this mean for the aerospace sector? From our point of view, the industry's well-established businesses must thoughtfully integrate both technological and financial opportunities to:

1. Improve the efficiency of the production system. Efficiency must be built on well-known lean production, Total Quality Management (TQM) concepts. But this needs to be combined with significant flexibility improvements in processes, complexity management of (international) supply chains/networks while preserving quality, and foremost keeping safety first and allowing for innovation. As in automotive industry, aviation has to cope with more and more product variants driven by customer demands, in particular with regard to cabin interiors. Industry 4.0 is about to improve the production systems and people capability, over a complex value network in order to deliver highest quality on time and on budget under more and more rapidly changing conditions. As a result, margins of all players in the system may increase. For German aviation industry, from OEM down to SME tier x suppliers, Industry 4.0 could offer an option to stay in the game successfully as stable technology, innovation and quality providers; and
2. Deliver appropriate products and services that provide added value to customers.

From the industry association's perspective, the task is to make any member, company, executive or entrepreneur—from OEM down to the suppliers of the smallest parts—aware of the strategic implications of this topic.

Following the results of a survey by pwc's Strategy&, German industry expects significant efficiency increase from Industry 4.0 employment. However, the percentage of companies expecting turnover increase is much lower. One way to interpret these results is to conclude that the strategic importance of these developments has not been fully understood. Moreover, German companies see numerous reasons why it would be difficult to implement elements of Industry 4.0, for example, inconclusive cost-benefit analyses, missing norms and standards, or distrust in Cloud applications.

Aviation Industry 4.0

It is apparent that the concept of Industry 4.0 needs to be defined individually for each industrial sector. While challenges might be similar, they need to be addressed with a sector-specific agenda. This allows industries to work on specific issues and to share the solutions with other sectors.

Distinctive characteristics of the aerospace industry include:

1. The sector manufactures highly complex, high-performance products which contain more than one million parts. While the customization degree is very high, fewer than 1000 units might be produced of any one aircraft type.
2. Aircraft programs have long lifecycles, up to 70 years from cradle to grave (10 years of development followed by 30 years of production plus a 30-year product life).
3. Driven by safety considerations, the sector is characterized by a highly regulated documentation and certification process imposed by national authorities. It is being harmonized at international level by LBA, FAA, CAAC, EASA, ICAO, etc.

These characteristics have led to a sector-specific product lifecycle:

1. In the aviation industry, technology management starts at earliest stages of the technology evolution, from invention to innovation. In aerospace small incremental improvements tend to have significant effects. For this reason, it is worthwhile to investigate potential technological improvements—even if the benefits are initially unclear.
   In addition, the development process must take incredibly high expectations in terms of safety, reliability and economic viability into account. All of this must essentially be done before the development phase is completed, which explains the large R&D budgets in our sector.
2. The aircraft is the world's most complex industrial product. A sophisticated division of labor between companies enables the industry to shoulder the high risks and investments required to bring a new product to market. Consequently, the development phase up to the point of certification is extremely complex.
3. Looking at the production phase, on one hand an aircraft entails over a million parts, but on the other hand, only a small number of units might be produced. Moreover, these units are highly customized based on customer specifications. Managing such a complex and deep supply chain remains challenging.
4. Finally, aircraft in service fall under a strict maintenance and overhaul regime to maintain their airworthiness. Countless unexpected events may occur during the lifetime of an aircraft, which requires the manufacturer to provide wide-ranging support over many years.

Due to these peculiarities, the aerospace industry often leads the way for other industrial sectors. Whether this will continue to be the case with the implementation of Industry 4.0 remains to be seen. It is clear, however, that there are a number of areas in which our sector could be a frontrunner for other industrial sectors:

1. The development phase
   Taking the next steps will require interconnected systems to virtually simulate both the product and the production process during the development phase. Such progress calls for a close cooperation between various research institutes; these alliances could eventually lead to a "consistent product and production data

model for international multi-tier cooperative product development." (Key word: Digital Mock Up; DMU).

Of growing importance for aeronautics are embedded systems and software engineering tools. The demands on software used in aerospace are much higher today than in the past, and we expect the complexities to increase exponentially. This has implications for software designers and programmers who are faced with increased complexity in software architectures, testing and system integration. For companies supplying avionics products, this means more demands on their software, higher risks and increasing costs (BDLI 2017). In view of this pioneering work, it is fair to say that aerospace provides the foundations which could be of interest to other industries, for instance, for autonomous transportation on road and rail, or manufacturing systems engineering.

2. The production phase

    Aerospace production circles already have a great deal of interest in implementing augmented reality applications for workers. Such aids provide pathways to partial industrial automation for high-end products such as aircraft which could later be used for other products being manufactured in small batches.

    A key objective for the aerospace industry remains: to set up a fully visible, manageable supply chain or network for its complex, customized products. This is to address potential supply chain disruption which must adapt to the challenges of higher production rates. This would benefit all participants in the supply chain, many of which are faced with varying levels of capacity utilization, missed delivery deadlines and full order books (Airbus 2015).

3. The maintenance, repair and overhaul (MRO) phase

    Although aircraft and engines already generate large data quantities, the industry is only now beginning to systematically use this treasure trove. Engine manufacturers, for instance, are working towards optimized maintenance intervals for individual engines. We expect such developments to yield new business models in the MRO sector. Moreover, advances in sensor technology and information transfer may develop into a "nervous system" for aircraft which in turn could form the basis for improvements in many areas, from structural health to passenger comfort. This needs to integrate big data technologies and appropriate data security.

5.2 Improving the Product: Technology Roadmap

Technology will be key for the German aerospace industry's future success, not only for the production process but also relating to products. Technology becomes even more critical once we appreciate the discrepancy—all other factors being equal—between the predicted doubling of air traffic (Airbus 2015) over the next two decades and the aviation industry's firm commitment to reduce its climate impact.

The obvious answer is to use technology to reduce emissions further and faster. The aviation and aerospace industries have an impressive track record in this respect, and our targets for the coming decades are no less ambitious. As part of the European Flightpath 2050 strategy, the aerospace industry has committed to develop technologies to reduce CO_2 and NO_x emissions by 75 % and 90 % respectively by 2050, compared with the baseline year 2000. Noise emissions are to decrease by a further 65 %. Since any incremental improvements in aerospace are difficult to achieve, the industry will require significant investments over the coming decades. National and European political support is therefore essential.

Another decisive factor is to ensure an effective coordination of the many organizations that form an EU-wide aerospace research network. The cooperative approach is already bearing fruit at various levels. Building on a strategic European research agenda, in 2015 BDLI and its members developed a more detailed Technology Roadmap. This roadmap encompasses all relevant areas of aerospace technology, including air traffic management, flight operations and flight control as well as all aspects of the aircraft itself such as configuration, engines, systems, cabin, materials and the production process. The objective is to provide valuable guidance to the existing national aerospace research network, as well as to the numerous supply chain companies which may not be directly involved in the process.

Our Technology Roadmap illustrates the interdependency of technological developments, but also makes clear that technologies can increasingly be introduced in existing models as soon as they become available (Fig. 4). Hence both evolutionary and revolutionary innovation will play a significant role over the coming decades. Airbus, for instance, does not plan to bring a completely new aircraft model to market before 2030, because the latest quantum leap in engine technology paves the way for significant efficiency gains of today's aircraft types. Jet engines and other aircraft components can be improved further still thanks to Additive Layer Manufacturing, new alloys and virtual design. In parallel, the industry is laying the foundations for revolutionary advances which could enable new aircraft configurations and propulsion technologies. Airbus Group's E-Fan has already proven that an electric aircraft can take to the skies. As a next step, the company plans to cooperate with Siemens to develop electric and hybrid propulsion systems at a new research center near Munich, and is already working with its partners on a possible architecture for a hybrid-powered regional aircraft with up to 90 seats (Fig. 5).

Fig. 4 Balance between continuous enhancement and new concepts

Outlook

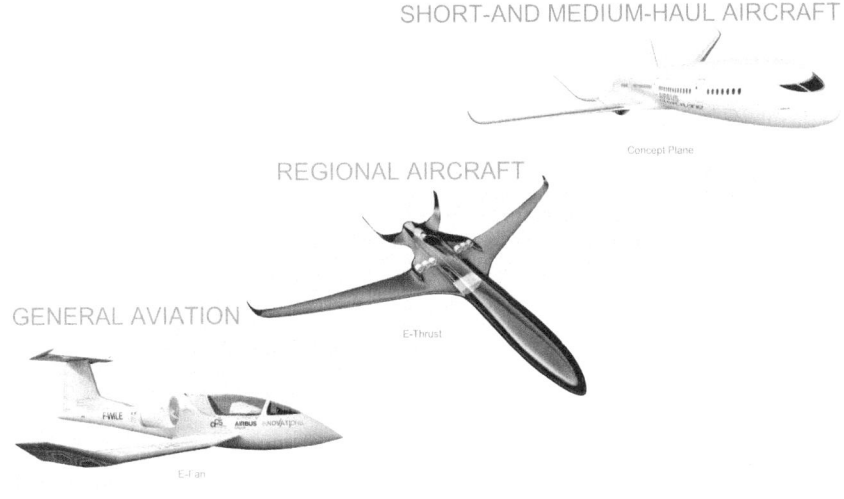

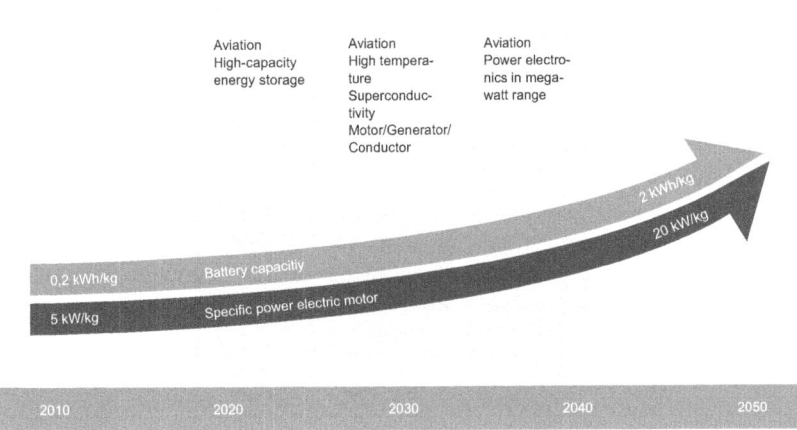

Fig. 5 Example—technology roadmap electric and hybrid drives

6 Preparing for the Future

We believe that every company in the sector is well advised to carefully consider the long-term strategic developments outlined in this chapter, and to prepare itself accordingly. Assuming that the manufacturers' standard scenarios become reality and growth persists in the coming decades, the aerospace industry will have to actively shape the future to safeguard its position as a global and European success story.

An association such as ours has a significant part to play in this process. Specifically, this means initiating and guiding crucial discussions, from the strategic make-up of the German aerospace industry to questions affecting small suppliers. Without fear or favor, we must bring both opportunities and risks to the attention of all stakeholders. To ensure that aerospace is recognized as a sustainable sector which contributes to the well-being of societies will require a high degree of coordination between people, companies and organizations. This is a major task for a sector which is organized as a network of companies and supply chains spanning the planet. As an association we contribute to this process on behalf of our members.

Its strategic relevance, long lead-times, and highly complex products render our sector unique. This makes us an ideal partner for decision-makers to discuss the importance of industrial policy, innovation, education and technology. High-profile events such as the Berlin Air Show, with its focus on new technologies and innovation, are excellent opportunities to showcase the sector's capabilities and its potential.

We hope that plans for a new aircraft generation to be developed over the next 20 years are a reminder that Europe should not leave the development of groundbreaking technologies to countries such as the USA or China. These are the very developments that attach a higher value to our sector's strategic importance. The supply chain is an important part of this debate, and should make its voice heard.

We are aware of the challenges ahead. At the same time we have reason to be optimistic. Aerospace remains one of the most exciting and successful industries of all. We connect people, nations and continents, provide the lifelines of the modern economy, help solve the most pressing problems of our time, and even enable humanity to venture beyond planet Earth. None of this would be possible without the thousands of companies in the supplier industry.

References

Airbus (2015) Global Market Forecast. Available via http://www.airbus.com/company/market/forecast/. Accessed 23 Jun 2016

Airbus (2016) Airbus exceeds targets in 2015. Press release. Available via www.airbus.com/presscentre/pressreleases/press-release-detail/detail/airbus-exceeds-targets-in-2015-delivers-the-most-aircraft-ever/. Accessed 5 Feb 2016

BDLI (2017) ESSEI—Das virtuelle Institut. Projekte, Forschung, Lehre, Netzwerk, Berlin
Boeing Commercial Airplanes (2015) Current market outlook 2015–2034. Available via www.boeing.com/resources/boeingdotcom/commercial/about-our-market/assets/downloads/Boeing_Current_Market_Outlook_2015.pdf. Accessed 23 Jun 2016
Koch V, Kuge S, Geissbauer R, Schrauf S (2015) Industry 4.0: opportunities and challenges of the industrial internet. Available via www.strategyand.pwc.com/reports/industry-4-0. Accessed 5 Feb 2016
Roland Berger Strategy Consultants, BDI (2015) Die Digitale Transformation der Industrie. Eine europäische Studie. Available via www.rolandberger.com/media/pdf/Roland_Berger_digital_transformation_of_industry_20150315.pdf. Accessed 5 Feb 2016
T-Systems (2016) Alles wird schlauer. Available via www.telekom.com/industrie-40. Accessed 5 Feb 2016
Wikipedia (2016) Computer-integrated manufacturing. Available via de.wikipedia.org/wiki/Computer-integrated_manufacturing. Accessed 5 Feb 2016

Stefan Berndes, Dr. was born in 1966. He is an aerospace engineer who graduated from Stuttgart University. He finished his academic career with a Ph.D. in philosophy of technology from Brandenburg Technical University Cottbus. He has more than 20 years of experience in project and innovation management, in particular in the aviation industry. Since 2005, he has been Head of the Department Air Transport, Equipment and Materials with the German Aerospace Industries Association (BDLI). In this capacity he has dedicated a lot of his efforts to strategic issues of aerospace supply chain development in Germany.

The manufacturer's authorised representative in the EU is Springer Nature Customer Service Centre GmbH, Europaplatz 3, 69115 Heidelberg, Germany. If you have any concerns regarding our products, please contact ProductSafety@springernature.com

Printed and bound by CPI Group (UK) Ltd, Croydon, CR0 4YY
23/03/2026
02076669-0002